TREATISE OF MAN
René Descartes

HARVARD MONOGRAPHS IN THE HISTORY
OF SCIENCE

HARVARD MONOGRAPHS
IN THE HISTORY OF SCIENCE

Chinese Alchemy: Preliminary Studies by Nathan Sivin
Leonhard Rauwolf: Sixteenth-Century Physician, Botanist, and Traveler by Karl H. Dannenfeldt
Reflexes and Motor Integration: Sherrington's Concept of Integrative Action by Judith P. Swazey
Atoms and Powers: An Essay on Newtonian Matter-Theory and the Development of Chemistry by Arnold Thackray
The Astrological History of Māshā'allāh by E. S. Kennedy and David Pingree
Treatise of Man: René Descartes French Text with Translation and Commentary by Thomas Steele Hall

Automated garden figures and main driving mechanism in the grottoes of the royal gardens at Saint-Germain-en-Laye. From engravings in Salomon de Caus, *Les raisons des forces mouvantes avec diverses machines tant utilles que plaisantes ausquelles sont adjoints plusieurs desseings de grotes et fontaines*, Frankfurt, J. Norton, 1615. This grotto is described by Descartes on page 13 of the French text.

XVIII

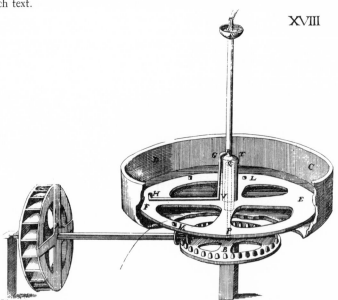

TREATISE OF MAN

René Descartes

French Text
with Translation and Commentary by

Thomas Steele Hall

Harvard University Press
Cambridge, Massachusetts
1972

AUTHOR'S NOTE

THE preparation of this edition of the *Treatise of Man* was begun in 1957 when I was in residence at Yale University as a National Science Foundation Faculty Fellow. The work was interrupted for a decade and resumed in 1967. Its completion owes much to numerous colleagues and helpful critics, among whom I am especially grateful to the late Professor John F. Fulton for initially encouraging this translation and commentary; to Professor I. Bernard Cohen for valuable editorial advice; to Suzanne Trocme, M.D., for checking the translation in its early stages; to Professor A. C. Crombie for reading and criticizing parts of the translation and notes; to Richard J. Durling of the National Library of Medicine for aid in the analysis of difficult classical and Renaissance texts; to Professor Phillip De Lacy and Dr. Margaret T. May, and to R. E. Siegel, M.D., for help in interpreting some significant issues of Galenic physiology; to Professor Isidore Silver for counsel on matters of syntax and interpretation; to Maurice Deixonne for assistance in solving orthographic questions; to Professor Saul Rosenzweig for leads to sources in the history of physiological psychology; to Dr. Audrey Davis for a critical reading of the manuscript in its later stages of development; to Professor Alexander Calandra and Richard Robb, M.D., for help with certain troublesome points in Cartesian optics; to Walter Pagel, M.D., for useful leads to medieval medical sources of Descartes; to Mrs. James J. Hanks, National Library of Medicine, for indispensable bibliographic assistance; to Robert J. Penella, a graduate research assistant, who cooperated in a very helpful way in the survey of pre-Cartesian physiological literature; to the staff of the Rare Book

Room, Countway Library, Harvard University Medical School, for numerous courtesies during two summers of research there; to Miss Shirley Funk for correcting, standardizing, and typing the manuscript; and to the National Science Foundation for financial support.

T. S. H.

CONTENTS

Foreword by J. Bernard Cohen   xi

Abbreviations   xvii

Introduction   xxiii

The Physiology of Descartes   xxvi

First French Edition: Synopsis of Contents   xxxiv

Bibliographic Materials
   Editions of the Treatise of Man   xliii
   Sources for the Study of Descartes's Physiology xlv

English Translation and Commentary   1

Facsimile of the First French Edition   119

Index   227

FOREWORD

DESPITE Descartes's outstanding importance in the development of science and medicine, his major writings are not for the most part available in well-annotated scholarly editions or translations, save for certain contributions to "pure" philosophy.[1] Furthermore, as may be seen by an examination of Sebba's *Bibliographia Cartesiana*,[2] we do not have available an extended series of major monographs on the actual contributions made by Descartes to the sciences: the psychology of perception, animal physiology, general medicine, the principles of motion, or — for that matter — physics[3] and astronomy in general. Even when treated in the literature, these topics tend to be presented without that specificity and analysis that give the reader the sense of satisfaction of having grasped more than the bare foundational elements of Cartesian science.

Mathematics and certain parts of physics have fared better than other parts of Descartes's science,[4] but even in these areas there is a want of translations and annotated texts. For example, "La

1. The great Adam and Tannery edition of Descartes's *Oeuvres* contains textual notes, but generally omits any historical or explanatory annotations.
2. G. Sebba, *Bibliographia Cartesiana: A Critical Guide to the Descartes Literature* . . . , The Hague, Nijhoff, 1964.
3. J. F. Scott's *The Scientific Work of René Descartes, 1596–1650*, London, Taylor and Francis, 1952, deals only with certain aspects of physics.
4. For principles of motion in Cartesian physics, see A. Koyré, *Études galiléennes*, Paris, Hermann, 1939 (repr. 1966), pt. 2, "La loi de la chute des corps, Descartes et Galilée"; and René Dugas, *La mécanique au XVIIe siècle*, Neuchâtel, Griffon, 1954. For Cartesian physics in general, see Paul Mouy, *Le développement de la physique cartésienne 1646–1712*, Paris, J. Vrin, 1934.

dioptrique" and "Les météores," two of the three tracts that were said by Descartes to be illustrative of his method, have only recently been rendered in English translation, in an edition wholly without commentary or annotations, and with somewhat misleading titles.[5] And the *Principia philosophiae*, the title of which was transformed by Isaac Newton into *Philosophiae naturalis principia mathematica*, and from which Newton and others learned the basic elements of Cartesian physics,[6] has never been translated into English, nor does it exist in a properly annotated scholarly edition in any language.

We are especially grateful, therefore, to have Descartes's *Traité de l'homme* made available to us in an English version, with a historical and analytical commentary. This work was written as one of two parts of a general *Traité du monde*, of which the other portion was a *Traité de la lumière*.[7] Although Descartes had spent some four years in preparing the two parts of this double treatise on "man" and "light," he withheld it from publication on learning that Galileo had been punished by the church for his advocacy of the Copernican doctrine of the motion of the earth in an annual orbit around the sun. *Le monde* was not published in the author's lifetime. We may be surprised to find Descartes suppressing both the physical and physiological parts of his treatise, thereby putting aside the fruits of protracted scientific study, including almost daily

5. René Descartes, *Discourse on Method: Optics, Geometry, and Meteorology*, translated, with an introduction, by P. J. Olscamp, Indianapolis, Ind., Bobbs-Merrill, 1965. The "Géométrie" had been translated earlier by D. E. Smith and M. L. Latham and published with a page-for-page facsimile of the first edition (1637); this edition was reprinted by Dover Publications, New York, in 1954. The *Discourse on Method* has been translated into English many times. The first English version was issued in London in 1649 ("Printed by *Thomas Newcombe*, and . . . to be sold at his house over against Baynards Castle") under the title, *A Discourse of a Method for the Well Guiding of Reason, and the Discovery of Truth in the Sciences;* a facsimile reprint was published in 1966 by Dawsons of Pall Mall, London.

6. On this point see A. Koyré, *Newtonian Studies*, Cambridge, Mass., Harvard University Press, 1965, III — "Newton and Descartes," pp. 53–200; also I. B. Cohen, " 'Quantum in se est': Newton's concept of inertia in relation to Descartes and Lucretius," *Notes and Records of the Royal Society of London*, 19 (1964), 131–155.

7. The two parts are generally treated today as if they were distinct, although Descartes himself held them to be inseparable. An Italian translation, *Il mondo, ovvero trattato della luce, traduzione di Gianfranco Cantelli*, Turin, Boringhieri, 1959, follows the posthumous edition of 1664 entitled *Le monde, ou traité de la lumière*. Details concerning the printing of the portion later entitled *Traité de l'homme* are given by Professor Hall in the present edition.

visits to the butchers' shops, where long and careful examinations of animal organs gave him a personal acquaintance with the fabric and functioning structure of the animal body. He also performed actual dissections on such particular organs as eyes and brains, and lungs and hearts. Yet, as Descartes wrote, he had to suppress the whole treatise (both the part we know today as the *Traité de l'homme* and that which we know as the *Traité de la lumière*), since he found the Copernican doctrine to be "so connected with every other part of my Treatise" that he "could not disconnect it without making the remainder faulty."[8]

The suppression of *Le monde* is often presented as an example of the far-reaching influence of the church on the intellectual life of the early seventeenth century, and this episode reinforces our views of how Galileo's trial deeply affected men of science in all Catholic countries. In the case of Descartes, however, the decision was all the more significant in the light of the energy spent in his actual study of the parts of animals. Since this activity is not generally known, let me quote a description of it from Baillet's biography, describing Descartes's activities in 1629 (based on the correspondence and here freely translated):[9]

In this conviction, he set about the execution of his design by studying anatomy, to which he devoted the whole of the winter that he spent in Amsterdam. To Father Mersenne he testified that his eagerness for knowledge of this subject had made him visit, almost daily, a butcher's, to witness the slaughter; and that he had caused to be brought thence to his dwelling whichever of the animals' organs he desired to dissect at greater leisure. He often did the same thing in other places where he stayed after that, finding nothing personally shameful, or unworthy of his position, in a practice that was innocent in itself and that could produce quite useful results. Thus he made fun of certain maleficent and envious persons who, intending to enjoy themselves at the expense of his reputation, had tried to make him out a criminal and had accused him

8. On this point see S. V. Keeling, *Descartes*, London, Oxford University Press, ed. 2, 1968, pp. 17–21.

9. A. Baillet, *La vie de Monsieur Des-Cartes*, Paris, D. Horthemels, 1691, pt. 1, pp. 196–197; this description may be documented by Descartes's letters, notably to Mersenne. An example of how little attention has been paid to this intense anatomical activity is given in the *Vie et oeuvres de Descartes* by Charles Adam, published as a "supplément à l'édition de Descartes," that is, the Adam and Tannery edition, Paris, L. Cerf, 1910, p. 124: "In Amsterdam, we know that he studied anatomy (perhaps made some dissections) with Plempius."

of "going through the villages to see pigs killed," although this was absolutely false so far as the villages were concerned. He read little at this time, to be sure, and he wrote even less. However, he did not neglect to look at what Vesalius and the most experienced of other authors had written about anatomy. But he taught himself in a much surer way by personally dissecting animals of different species; and he discovered directly many things more detailed than the ones that all these authors had reported in their books. For several years he continued this practice, diversifying his pursuits meanwhile through other studies. With such exactitude did he examine even the smallest parts of animal bodies that no professional physician could boast of having taken closer notice of these. He declared to Father Mersenne that after ten or eleven years of searching he had found nothing, however small, whose purpose, and whose formation through natural causes, he felt unable to explain in detail, just as he had explained the purpose and formation of a grain of salt or a tiny snowflake in his treatise "Les météores." Yet after an endless number of experiments and after many years' application to studies of this kind, he was too modest to think himself able to cure even so much as a fever. His long endeavor had acquainted him only with the animal in general, and animals are not subject to fever. It was this that obliged him to apply himself thereafter more closely to the study of man, for man is subject to fever.

In the Haldane biography,[10] regarding Baillet's statement that Descartes almost daily visited the butchers' shops, taking home the portions of the animals which he wished to dissect at leisure, it is said that "his biographer's defense sounds strangely," especially the reference to the accusation of going to villages to see pigs slaughtered, which "Baillet declares . . . false so far as the villages were concerned." Descartes no doubt earned ridicule and even contempt as a student who thus soiled his hands by direct contact with the structures of animals, and who sought truth in the experience of direct contact, rather than solely in books and disputations.

One of the reasons why so many different kinds of readers will find the *Traité de l'homme* to be of great interest is that in it there is no separation of physiology or even psychology from general physics and cosmology. The image of man presented to us by Descartes is, as Professor Hall points out, neither purely philosophical nor purely physiological, but is linked to the whole interrelated structure of Cartesian general science and philosophy, physics and

10. E. S. Haldane, *Descartes: His Life and Times*, London, Murray, 1905, pp. 126–127.

cosmology. Like other writings of Descartes, the *Traité de l'homme* is not limited to the confines of a single discipline. Of course, he lived in an age that had not yet seen the establishment of the narrow restrictions of specialties within the sciences, nor had the divorce of science itself from philosophy yet occurred. Thus Descartes's treatise on dioptrics is a classic, among other reasons, for its modern statement of the process whereby we perceive sizes, shapes, and distances; his treatise on the passions of the soul is numbered among the primary writings in the development of our understanding of mind and brain.[11]

Readers will find especially valuable Professor Hall's analysis of Descartes's *Traité de l'homme*, in the form of an extensive series of annotations together with an informative introduction. Professor Hall particularly directs our attention to the medical and physiological emphases of the work, and its imbalance (a stress on mechanical aspects of nervous function and on the cardiac cycle, and a neglect of the physiology of nutrition) — a not-unnatural outcome of the stress placed by Cartesian science on the doctrine of "matter and motion." This example shows how Cartesian physiology takes us at once to the general principles of the Cartesian "mechanistic philosophy," itself so important for the development of science and philosophy in general throughout the seventeenth century. One has only to flip through the pages of Descartes's treatise, and Professor Hall's notes, to see at once that the *Traité de l'homme* is an admirable instance of the inseparability of Cartesian science and philosophy, Cartesian medicine and physics, and Cartesian psychology and cosmology.[12]

The importance given to this early work by its author may be seen from the very fact that the original title was *Le monde*, of which it formed the second part. *Le monde* conveys to the reader the author's intention to develop a system of cosmology founded on a

11. See, for example, A *Source Book in the History of Psychology*, R. J. Herrnstein and E. G. Boring, eds., Cambridge, Mass., Harvard University Press, 1965.

12. The interactions between Cartesian philosophy and science, stressed by Professor Hall, provide an essential key to our understanding of Descartes's contributions to physiology and psychology, as well as to our understanding of his work in physics and mathematics. See L. C. Rosenfield, *From Beast-Machine to Man-Machine* . . . , New York, Oxford University Press, 1941 (enl. ed., New York, Octagon Books, 1968); and A. Georges-Berthier, "Le mécanisme cartésien et la physiologie au 17ème siècle," *Isis*, 2 (1914), 37–89 and 3 (1920), 21–58.

physics so universal that it could apply to animal functions and to man himself, as well as to inanimate objects — to the world as a whole and all it contains.

For these many reasons, we are particularly pleased that the Harvard Monographs in the History of Science may include an edition of so significant a work, in the annotated translation made by Professor Hall,[13] equally distinguished for his work in zoology and in the history of animal biology and physiology. This new version of Descartes's *Treatise of Man* should attract the attention of a variety of readers with rather different scholarly orientations, including the growth of philosophy, the intellectual history of the seventeenth century, the development of modern physiology, and the rise of scientific psychology.

I. BERNARD COHEN

13. Descartes's *Traité de l'homme* has never before been translated into English in its entirety; nor have any significant extracts been included in volumes of selections in English. No part is to be found, for example, in the widely used two-volume set of the philosophical works of Descartes translated by E. S. Haldane and G. R. T. Ross, nor does a portion occur in the volume of philosophical writings translated and edited by E. Anscombe and P. T. Geach. This work is also conspicuously missing from N. Kemp Smith's book of Descartes's philosophical writings and the volume of selections edited by R. M. Eaton. A short excerpt, however, does appear in the source book cited in note 11.

An annotated translation into German appears as *René Descartes: Über den Menschen (1632), sowie Beschreibung des menschlichen Körpers (1648): Nach der ersten französischen Ausgabe von 1664 übersetzt und mit einer historischen Einleitung und Anmerkungen versehen, von Karl E. Rothschuh* (Heidelberg, Lambert-Schneider, 1969). An Italian version, companion to *Il mondo* (see note 7), was published under the title, *L'Uomo: Introduzione e traduzione di Gianfranco Cantelli*, Turin, Boringhiere, 1960.

ABBREVIATIONS OF REFERENCE WORKS CITED IN INTRODUCTION AND NOTES[1]

Aristotle
  de An.   *De anima*
  GC   *De generatione et corruptione*
  Juv.   *De juventute*
  Long.   *De longitudine et brevitate vitae*
  MA   *De motu animalium*
  Mem.   *De memoria*
  PA   *De partibus animalium*
  Resp.   *De respiratione*

AT   Charles Ernest Adam and Paul Tannery, eds., *Oeuvres de Descartes*, 11 vols., Paris, L. Cerf, 1897 to 1910, plus vol. 12, "Vie et oeuvres de Descartes," plus suppl., "Index général," 1913. Republished, 12 vols., Paris, J. Vrin, 1956 to 1957 and 1964 to 1967

Bartholin
  Enchiridion   Caspar Bartholin, *Enchiridion physicum ex priscis & recentioribus philosophiis* . . . , Strasbourg, L. Zetzner, 1625 (first publ. Frankfurt, 1610)
  Institutiones   *Anatomicae institutiones corporis humani* . . . , Strasbourg, C. Scher, 1626 (first publ. Wittenberg, 1611)

---

1. This list comprises works referred to with some frequency; seldom-cited works are given in unabbreviated form in the text.

Bauhin
   Theatrum   Caspar Bauhin, *Theatrum anatomicum novis figuris aeneis illustratum* . . . , Frankfurt, M. Becker, 1605 (first publ. Basel, 1590, as *De corporis humani fabrica* . . .)

Columbus
   Anatomica   Realdo Colombo, *De re anatomica libri XVI*, Venice, N. Bevilacqua, 1559

Crooke
   Mikrokosmographia   Helkiah Crooke, *Mikrokosmographia* . . . , London, T. and R. Cotes, 1631 (first publ. London, 1615)

Descartes
   Anatomical
   Excerpts   Written at various times between 1631 and 1648; first published posthumously by Foucher de Careil in *Oeuvres inédites de Descartes* . . . , Paris, A. Durand, 1859 to 1860 (see AT 10:207 and AT 11:545–547)

   de Homine   Abbreviation used in citing original Latin edition of *Treatise of Man* (see list of editions of this work)

   Description
   of the Body   *La description du corps humain*, first publ. 1664 with Clerselier's first French edition of *Treatise of Man*

   Dioptrics   "La dioptrique," the first of three essays published with the *Discourse* as an introduction to them (the others, "Le météores" and "La géométrie"), all three being intended as "trials" (*essaies*) of the analytical method developed in the *Discourse*

   Discourse   *Discours de la méthode pour bien conduire sa raison* . . . , published anonymously, Leiden, J. Maire, 1637

   Generation   "Primae cogitationes circa generationem animalium," first publ. in *Opuscula posthuma physica et mathematica* . . . , Amsterdam, P. and J. Blaeu, 1701

l'Homme	Abbreviation used in citing first French edition of *Treatise of Man* (see list of editions of this work)
Light	*Le monde de M*^r. *Descartes, ou le Traitté de la lumière et des autres objets des sens* . . . , Paris, N. Bobin and J. Le Gras, 1664
Man	*L'homme de René Descartes* . . . , Paris, issued separately (but from the same plates) by C. Angot, N. Le Gras, and T. Girard, 1664 (see list of editions of this work)
Meditations	*Meditationes de prima philosophia* . . . , Paris, M. Soly, 1641; translated by "M. le D.D.L.N.S.," (the Duc de Luynes) *Les méditations métaphysiques* . . . , Paris, J. Camusat and P. Le Petit, 1647
Meteors	"Les météores," one of three treatises published with the *Discourse*
Passions	*Les passions de l'âme*, Amsterdam, L. Elzevir, 1649, also sold in Paris under the imprint of H. Le Gras
Principles	*Principia philosophiae*, Amsterdam, L. Elzevir, 1644; translated "by one of our friends," (l'Abbé Picot), *Les principes de la philosophie*, Paris, H. Le Gras, 1647. References will cite both the Latin original (AT 8:1–348) and the French translation (AT 9, pt. 2:1–352)
Regulae	"Regulae ad directionem ingenii," first published from a copy of the original ms. in *Opuscula posthuma, physica & mathematica*, Amsterdam, P. and J. Blaeu, 1701
Fabricius *Visione*	Hieronymus Fabricius ab Aquapendente, *De visione, voce, auditu*, Venice, F. Bolzetta, 1600
Falloppius *Opera*	Gabriello Falloppio, *Opera omnia in unum congesta* . . . , Frankfurt, C. Marnius and J. Aubrius, 1600 (earlier editions, Venice and Frankfurt, 1584)

Fernel
  Physiologie

Jean Fernel, *Les VII. livres de la physiologie
. . . traduits en françois par Charles de Saint
German*, Paris, J. Guignard, 1655 (a translation
of the physiological section of Fernel's *Medi-
cina*, Paris, 1554, this section being a revision
and simplification of *De naturali parte medicinae libri septem*, Paris, 1542)

Galen
  Opera

Galenus, *Opera omnia*, see below, abbreviation
"K"

  Anat. Adm.   *De anatomicis administrationibus*
  de Alim fac.   *De alimentarum facultatibus*
  de Caus. resp.   *De causis respirationis*
  de Loc. aff.   *De locis affectis*
  de Mot. mus.   *De motu musculorum*
  de Plac.   *De placitis Hippocratis et Platonis*
  de San.   *De sanitate tuenda*
  de Simp. med.   *De simplicium medicamentorum temperamen-
                        tis et facultatibus*
  Meth. med.   *De methodo mendendi*
  Nat. fac.   *De naturalibus facultatibus*
  Sympt.   *De symptomatum causis*
  Temp.   *De temperamentis*
  UP   *De usu partium corporis humani*
  UR   *De utilitate respirationis*

Gilson
  ISC

Étienne Gilson, *Index scholastico-cartésien*,
Paris, F. Alcan, 1912, 1913; New York, B.
Franklin, 1964

Harvey
  *Circulation
  of the Blood*

William Harvey, *Exercitationes duae anato-
micae de circulatione sanguinis, ad Johannum
Riolanum filium . . .* , Rotterdam, 1649, trans.
K. J. Franklin, *The Circulation of the Blood
. . .* , Oxford, Blackwell, and Springfield, C.
Thomas, 1958

Motion of the Heart	*Exercitatio anatomica de motu cordis et sanguinis in animalibus* . . . , Frankfurt, 1628, trans. K. J. Franklin, *Movement of the Heart and Blood in Animals* . . . , Oxford, Blackwell, and Springfield, C. Thomas, 1957
K	Karl Gottlob Kühn, *Medicorum graecorum opera quae extant*, vols. 1 to 20, Leipzig, Cnobloch, 1821 to 1833, cited by volume and page number (for example, K3, 265 for Kühn, vol. 3, p. 265)
Kepler *Werke*	Johann Kepler, *Gesammelte Werke herausgegeben im Auflag der deutschen Forschungsgemeinschaft* . . . , W. von Dyck and M. Caspar, eds., Munich, C. H. Beck, 1938 to date
Dioptrics	*Dioptrice seu demonstratio eorum quae visui* . . . , Augsburg, D. Francus, 1611
du Laurens *Oeuvres*	Andreas Laurentius, *Toutes les oeuvres de* . . . , Rouen, R. du Petit Val, 1621 (first publ. Paris, 1600)
Historia	*Historia anatomica humani corporis* . . . , Frankfurt, M. Becker, 1600
May *Galen*	Margaret T. May, *Galen on the Usefulness of the Parts of the Body* . . . , Ithaca, Cornell University Press, 1968
Paré *Oeuvres*	Ambroise Paré, *Les oeuvres de* . . . , Paris, G. Buon, 1575, 1579, 1585 (date of edition specified with each citation)
Anatomie	*Anatomie universelle du corps humain* . . . , Paris, le Royer, 1561
Piccolhomini *Praelectiones*	Archangelo Piccolhomini, *Anatomicae praelectiones* . . . , *explicantes mirificam corporis*

humani fabricam . . . , Rome, B. Bonfadinus, 1586

Riolan
   Oeuvres

Jean Riolan (the younger), *Les oeuvres anato-miques de M. Jean Riolan . . . mis en françois par M. Pierre Constant*, Paris, D. Moreau, 1629

   Anthropographie

"De l'anthropographie" in *Oeuvres*, above

Rothschuh
   D.

K. E. Rothschuh, *René Descartes: Über den Menschen (1632) sowie Beschreibung des menschlichen Körpers (1648)* . . . , Heidelberg, Lambert-Schneider, 1969

Scheiner
   Oculus

Christopher Scheiner, *Oculus: hoc est, funda-mentum opticum* . . . , Innsbruck, Agricola, 1619

Siegel
   Galen's System

Rudolph E. Siegel, *Galen's System of Physiol-ogy and Medicine* . . . , Basel and New York, S. Karger, 1968

Vesalius
   Fabrica

Andreas Vesalius, *De humani corporis fabrica libri septem*, Basel, J. Oporinus, 1543

INTRODUCTION

O N July 18, 1629, René Descartes wrote to the eminent Oratorian priest, Guillaume Gibieuf, about "a little treatise which I am starting" and which "I hope to finish in two or three years and perhaps after that will decide to burn . . . ; for if I am not clever enough to do something well, I shall at least try to be wise enough not to publish my imperfections" (AT 1:17). Less than three years later in a letter to his constant correspondent, Marin Mersenne, he reported progress on the treatise and said it contained "a general description of the stars, the heavens, and the earth," as well as bodies on the earth. Concerning the latter, he hoped to "show the way to an understanding of them through the combined use of experiment and reason" (April 5, 1632, AT 1:243). Descartes entitled his treatise *The World*.

Only a few months thereafter, Descartes wrote again to Father Mersenne. "In my *World*," he said, "I shall speak somewhat more of man than I had thought to before, because I shall try to explain all his principal functions. I have already written about those that pertain to life, such as the digestion of food, the beating of the pulse, the distribution of nutrients etc., and the five senses. Now I am dissecting the heads of different animals in order to explain what imagination, memory, etc. consist of. I have seen the book [Harvey's] *De motu cordis* of which you spoke to me earlier, and find I differ a little from his opinion which I saw only after having finished writing about this matter" (AT 1:263).

Of the entire work projected by Descartes under the title *The World*, two substantial portions survive, which he called *Treatise of*

Light and *Treatise of Man*. They were written in French. There is unassailable evidence that a passage of uncertain length and contents, connecting the two, is missing (see AT 11:iii–iv). A third part, which would probably have borne the title *Of the Soul*, was planned and possibly drafted. It may have been destroyed by Descartes himself. In any case, it was not among his effects when he died (see Rothschuh, D., 135).

Neither of the two extant parts was published during the lifetime of the author. Descartes was influenced to withhold them, partly by Galileo's difficulties with the Inquisition, but he did include a paraphrase of parts of the *Treatise of Man* in section V of his *Discourse on Method*, which was published anonymously in 1637. It was not until twelve years after the death of Descartes that the *Treatise of Man* itself first appeared — and then in an imperfect Latin translation by Florentius Schuyl (*Renatus Des Cartes, De homine figuris et latinitate donatus*, Leyden, apud Franciscum Moyardum et Petrum Leffen, 1662). Two years thereafter, Descartes's literary heir and executor Claude Clerselier published French versions of the two parts of *The World* separately, the first as *Le monde de M^r. Descartes, ou le Traitté de la lumière* (Paris, Jacques Le Gras, 1664) and the second as *L'homme de René Descartes* (Paris, 1664).

The manuscripts Clerselier used for his editions of the two treatises, probably autographs, were among personal papers of Descartes discovered in Stockholm, where he died in 1650 after a brief winter sojourn. Many of the Stockholm papers, *Man* among them, were sent by sea to Rouen and thence in a smaller vessel to Paris. They survived immersion when the latter vessel sank in a storm. Nothing certain is known about the history of the manuscript after its publication. Clerselier died in 1684, bequeathing his Cartesian papers to abbot J-B. Legrand who turned them over, in time, to Descartes's biographer Baillet; but whether the text of *l'Homme* was still part of the collection remains a mystery (AT 1:xv–xix).

Official permission to publish *l'Homme* was accorded to editor Clerselier, but he tells us in a footnote to his preface that he "ceded and conveyed the privilege to Charles Angot, to Jacques and

Nicolas Le Gras, and to Théodore Girard, book merchants of Paris." Of the imprints issued by these three firms all were made from the same set of plates. Even the three title pages were printed from the same plate of type into which each issuer inserted his own insignia, name, and place of business.

THE PHYSIOLOGY
OF DESCARTES

A T the beginning of the sixteenth century, physiology was still dominated by Greek assumptions and axioms, but these had been modified by thinkers who had variously combined them with elements drawn from alchemy, astrology, magic, logic, and the doctrines of the church. From about the time of Vesalius (1542), the modified Greek ideas began to suffer erosion. Certain late sixteenth- and early seventeenth-century medical thinkers outspokenly challenged the Hellenic tradition. A majority accepted it — but tended to evoke it less and less often so that it was weakened, in part, through neglect.

Historical Position of Descartes's Physiology

Our commentary will show that Descartes was thoroughly conversant with theoretical developments in the physiology of the period immediately preceding his own. He saw in these developments a challenge to the legacy of Greece. He wanted to articulate that challenge and to make it decisive by replacing Greek ideas with ideas of his own. His success in achieving this goal was substantial but, as we shall discover, not total. In the physiological system that he developed, many traditional elements remained.

To assess the results, we must make a distinction between the detailed accounts of body function that Descartes provided and the broad premises that seemed — to him at least — to dictate these details. Here a paradox presents itself. The detailed explanations he developed were in large part promptly rejected, whereas his general premises eventually prevailed. Unfortunately for his reputation as a physiologist, he applied his broad assumptions with a kind of deductive blindness, a disregard for verification. As a result, his specific conclusions were fated to find acceptance with only a small circle of admirers.

Yet his influence was beneficial, decisive even, since in the hands of other more inductively and empirically oriented inquirers, his general assumptions produced more lastingly accepted results. Thus the influence of Descartes was felt at first on the broad conceptual level — through the urgent application of his materialistic and mechanistic ideas — and only later, and through the work of others, on the level of concrete detail.

But what were the modified Greek premises, the physiological "isms," that seemed to Descartes to require revision or even rejection and replacement? Among them, five were perhaps most important:

(a) *Humoralism: the "Hippocratic" assumption that the body comprises four humors, whose different blends or temperaments are responsible for the functions of the different organs.* Galen had altered this doctrine to make the humors nutrient precursors, but not persistent components, of the tissues. Descartes proceeded for the most part without evoking the doctrine of humors at all. Instead, he explained everything that happens in the body in terms of his theory of corpuscular physics.

(b) *Dualism of body and spirit: the Stoic (and alchemical) idea that all things, including living things, consist of both body (soma) and spirit (pneuma, or often, in alchemical thought, "quintessence").* This particular kind of dualism had lost much of its force in physiology by the time of Descartes, having had relatively few supporters after its forceful advocacy by Paracelsus (d. 1541). Nevertheless, there were still a few subscribers to the doctrine up to and into the seventeenth century, most notably Francis Bacon (in his *History of Life and Death*, 1623). In the writings of Descartes, we hear nothing about quintessential spirits; they find no place in a mechanical-material scheme of the kind he proposed.

(c) *Pneumatism: the neo-Galenic assumption of three spirits or pneumata — animal, vital, and natural.* These spirits were variously conceived, but were ordinarily regarded as intermediating between the body and the soul. The soul, in turn, was viewed as the ultimate cause of all physiological functions. Galen had definitely posited vital and animal spirits but had been somewhat noncommittal about the natural variety. In medieval and pre-Cartesian Renaissance pneumatology, three spirits were the rule. Of the three, Descartes retained only one, the animal kind. He assigned these animal spirits a variety of roles in the brain, as we shall see, and in the nerves, where he made them responsible for motor transmission.

(d) *Interpretation in terms of "faculties": the Galenic assumption of dispositive conditions or "faculties" that characterize the body as a whole as well as the individual organs.* Ideas about faculties differed. In general they were believed to permit in some manner, all the operations of the organs. Galen acknowledged that "faculty" is merely a name we apply when we are ignorant of those latent arrangements in a body-part which permit its overt or patent operations (*Nat. fac.*, bk. 1, chap. 3).* Descartes's objective was, precisely, to show what these latent factors are. He explicitly rejected Galen's faculty theory and substituted mechanical alternatives of his own.

(e) *Animism, or psychism: (as we shall use the term here) the attribution of all the patent manifestations of life to a latent causal entity, the psyche or life-soul.* This idea was held, in various forms, by all Greek and by most medieval and early Renaissance thinkers. Descartes did retain the soul, but only as an agent of thought — of volition, conscious perception, memory, imagination, and reason. All other physiological functions, including nerve-mediated ones, he attributed to the movements of material corpuscles and especially to the heat which he, and almost every physiologist of the day, believed to be generated in and distributed by the heart. And heat too, for Descartes, was corpuscular motion.

If the foregoing notes make Descartes appear primarily an iconoclast who delivered the coup de grace to already crumbling ideas,

* For more about Galen's faculty theory, see T. S. Hall, "On biological analogs of Newtonian paradigms," *Philosophy of Science,* 35 (1968), 9–10.

that impression is partly correct. But he went further and substituted for the crumbling edifice a new structure partly his own. His view of the body is correctly termed "mechanistic," provided we understand what Cartesian mechanics entailed. Its major assumption is that the physical properties of bodies are reducible to the properties of three elements or variants of matter — specifically, to the shapes and motions with which the particles of the elements had originally been endowed. Like other physicists of his era, Descartes reasoned mechanistically about objects varying in size from the invisibly minute to the celestial. Regardless of the size of the object, the analysis brought Descartes, in the end, to the particles of which he believed the object to be composed. The swirling vortexes of the cosmos — and the water droplets of the rainbow — were, for Descartes, corpuscular machines.

As for man's body, Descartes sometimes likened it to those machines or mechanical contrivances represented by clocks, or automated fountain-figures, or systems of pulleys and levers. Such, however, was not the primary sense in which he considered man a machine. His physiological essays, including Man, make it clear on page after page that to him the mechanics of the physiologist is the mechanics of very small things. It begins just below the level of vision and reaches downward to the level of elementary particles of matter.

It is interesting in this connection to compare the mechanical views of Descartes with those of his great contemporary, William Harvey. The latter's mechanical interpretations were numerical and metrical; he measured the output of the heart and counted its beats per minute. In the physiology of Descartes we encounter no numbers. His procedure is mathematical only in the sense of having geometry as its model, that is, in reasoning deductively from a few self-evident axioms.

If we look retrospectively at these developments it seems clear that, in order to progress as it did, physiology needed the metrical-numerical mechanics of Harvey. It needed also his experimental genius (Descartes's few physiological experiments were not in a class with Harvey's). Despite the success of his method, however, Harvey shows a lingering acceptance of teleologic, anti-materialistic, and even animistic assumptions of the kind Descartes deplored.

Physiology also needed its Descartes. It needed the freedom he proclaimed from animistic, humoral, pneumatic, and "facultative" working ideas and the relentless application of the materialistic and mechanistic ideas he espoused.

Relation of Descartes's Physiology to His Other Endeavors

The life achievement of Descartes was his construction of a philosophical system embracing cosmology, physics, mathematics, epistemology, psychology, and certain aspects of religion. These elements did not act independently in his system but were woven together to form an integrated whole. Singly and in combination, moreover, they influenced his approach to physiology when he turned his attention there.

The link with cosmology appears, especially, in Descartes's response to one of the oldest and most central of all investigative problems, that of man's relation to the cosmos. It is not insignificant that Descartes envisioned his *Man* as part of a more comprehensive work to be entitled *The World*. Physiologists of all previous eras had drawn analogies between man, to whom they often referred as a "little universe" or microcosmos, and the universe at large, or cosmos. Descartes's way of making this connection was to subject both microcosmos and cosmos to a single set of divinely instituted decrees. And it was precisely in so doing that he forged connections between cosmology, physiology, and physics. For in constituting the human body of elementary particles, he used the same three basic kinds that he used in constituting the cosmos. He portrayed these particles as acting everywhere in obedience to uniform physical laws: laws of motion instituted by God when He created the world.

In Descartes's own view it was mathematics (which he coupled in a particular way with his theory of knowledge) that permitted him to draw these conclusions. He believed that as in geometry, so in every aspect of the search for knowledge, not excluding religion, one must begin by abandoning preconceptions. Starting anew and accepting the least possible number of completely self-evident axioms, one must build an inferential — but by its nature indubitable — image of man. Through mathematics thus conceived one might solve all sorts of problems, even medical problems, including that of aging. He even essayed, in the *Discourse on Method*, a "mathematical" (in other words, a rational and deductive) demonstration of the existence of God.

It should be noted, finally, that the human image built in this inferential way by Descartes had a strong pyschological component, and that he linked his psychological theory with his theories of cosmology, physiology, and physics. This is not the place to consider his subdivision of the cosmos into two entities — one space-filling (*res extensa,* body or matter) and the other conscious or "cogitant" (*res cogitans,* soul or mind) — except to say that as we proceed we shall hear much about both the independent and the interdependent activities of these two entities at their meeting place, the nervous system of man. This system was, for Descartes, an *extended entity*, a corpuscular mechanism arranged to act, in most of our behaviors, reflexly or independently of the *cogitant entity*, mind — but also able to interact with mind when circumstances required.

Derivative Nature of Descartes's Physiology

The commentary in this first English edition of the *Treatise of Man* is offered with three objects in view: first, to interpret passages that are not intrinsically clear; second, to suggest what Descartes had to say about the human body in treatises other than *Man* and in letters and unpublished fragments; third, and most important, to emphasize a central point about Descartes's biological endeavor, namely its highly derivative nature.

The derivativeness of Descartes's explanations is significant in the light of his expressed intention to discard traditional ideas and build a new image of man by proceeding deductively, as in geometry, from self-evident axioms (*idées claires et nettes, notions primitives*) (see, for example, his letter to Elizabeth, AT 3:665–666). The actual result is far from the indicated purpose. *What Descartes said about the body can only be correctly assessed in terms of Greek and Renaissance thories which he borrowed, or built upon, in arriving at theories of his own.*

From his correspondence and from other contemporary evidence we know that Descartes was familiar with certain Hippocratic and

Galenic treatises, with Scholastic writers on medicine including commentators on Platonic and Aristotelian biology, and with certain major biological writers of the mid-sixteenth to early seventeenth centuries. He specifically cites the younger Riolan, Bauhin, Harvey, Fabricius, Aselli, and Bartholin. In turn, these authors cite, and often discuss in detail, the ideas of other writers not named by Descartes (this is especially true of Bauhin). As to authors not mentioned by name, Georges-Berthier, who until very recently had done the best scholarly work on Cartesian biology, thought it "improbable that Descartes was unacquainted with some writings of van Helmont, Sylvius, Fallopius, Columbus, and Fernel" (see list of secondary sources: Georges-Berthier, 1914, 44). On circumstantial evidence, it appears that the name of Piccolhomini should be added to those listed by Georges-Berthier.

Under the pretext of building a wholly new physiology, Descartes makes very few textual allusions to authors from his past. Yet their influence is apparent on page after page of his works. It appears in his residual adherence to certain traditional errors (such as the latest notions about animal spirits and about the putative heat of the heart) as well as in his adoption without acknowledgment of largely correct ideas set forth by his immediate predecessors (for example, Keplerian optics). We do not mean that Descartes laid false claims to other people's ideas; but in his published works he passed over the authorship of many such ideas in silence.

This situation defines the task of the historian seeking to identify sources. Only occasionally can we say that Descartes first grasped a particular idea while reading a particular book or letter from a correspondent or friend. More often we must depend — as did Gilson in his search for the philosophical sources of Descartes — on the circumstantial but transparent evidence contained in what Descartes had to say. From such evidence we learn that he was extensively versed in, and built his system upon, the physiological and medical traditions of the period immediately preceding his own. It is that tradition, insofar as it is relevant, that our footnotes will endeavor to sketch.

Of Descartes's predecessors, it will be Galen whose name and whose ideas will appear most frequently in our notes. That this should be so is inevitable for several reasons. Authorities agree that

Descartes was directly acquainted with the major Galenic texts which formed the primary basis of Renaissance medical thought. In many cases, to be sure, the ideas developed by Descartes take as their points of departure the observations and ideas of his own immediate forerunners and contemporaries. We shall hear much about these thinkers and their influence on our author. But even their opinions, and explanations of the body, are scarcely comprehensible without some understanding of Galen, because it was Galenic medicine which in a variety of ways the late sixteenth- and early seventeenth-century investigators developed, or departed from, or denied.

Galen, however, was not the only Greek source of Renaissance medical thought. The Scholastic tradition had made sixteenth- and early seventeenth-century thinkers conversant with Plato, especially the *Timaeus,* and with the major and minor biological and psychological treatises of Aristotle. These Descartes must have known in part directly and, in the case of Aristotle, through the commentaries, especially those of the commentators at Coimbra, whose works were read in the Jesuit college Descartes attended. The influence of the commentators on Descartes has been extensively explored by Gilson, whose *Index scholastico-cartésien* contributed immeasurably to the image of Descartes that our comments will try to convey.

In addition to Plato and Aristotle, other pre-Galenic Greek thinkers, the Hippocratic authors, the Stoics, and the Alexandrians touched Descartes. In some cases this influence was direct, but perhaps primarily it was felt through the representation of these scholars by Galen, who was a medical historian as well as a practitioner, experimenter, and author.

FIRST FRENCH EDITION: SYNOPSIS OF CONTENTS

C LERSELIER'S first edition of the original French version contains in addtion to the text of *l'Homme* a number of other items. The total contents of the volume are suggested in the following outline.

(1) Clerselier's dedicatory *Epistle to Monseigneur de Colbert* (five unnumbered pages)

With a certain irony, but also with a certain grace, Clerselier contrives a comparison between Minister de Colbert (who through attention to budgetary details brings financial order to the realm) and philosopher Descartes (who through attention to physical details brings order to man's picture of the world).

(2) Clerselier's *Preface* (sixty-three unnumbered pages)

Clerselier devotes much of his preface to problems of preparing the manuscript for publication. He expresses discontent with having been anticipated by Schuyl, whose Latin translation had appeared two years before. If Schuyl had "done me the favor of informing me [of his plans], . . ." Clerselier says, "I could have prevented his falling into . . . pitfalls inevitable in view of the defective copy he used." As for Schuyl's illustrations, Clerselier considers them well engraved and well printed (a fair judgment) but "uneven in intelligibility and ill-suited to an understanding of the text." He properly praises Schuyl, however, for his Latin preface (see item 6 below).

Clerselier advises us that he contracted with Gerard van Gutschoven and Louis de La Forge, professors respectively at Louvain and La Flèche, to make separate complete sets of illustrations as called for by the text. Wherever the two illustrators, working independently, made substantially similar drawings, Gutschoven's were chosen as clearer. Where the two illustrators' drawings of the same subject differed considerably, Clerselier published both, representing van Gutschoven's by the letter "G" and La Forge's by "F". Clerselier also discovered among his legacy of Cartesian manuscripts two figures roughly drafted by Descartes himself — one of which, on muscle antagonism, he published on page 17 of his edition. This Clerselier labeled "D." (In the present edition all figures are reproduced with the French text exactly as they appear in the Clerselier volume; when used with the English translation, however, some have been reduced in size or have undergone minor alterations in the letter designations for the convenience of the reader.)

In addition to discussing the manuscripts and especially the illustrations, Clerselier quotes and discusses some passages from St. Augustine concerning the separateness of body and soul. His purpose here is to safeguard the reputation of Descartes by showing the propriety of the distinction between the cognitive aspect of man (*res cogitans*, soul or mind) and his corporeal aspect (*res extensa*, body or matter).

(3) Text of the *Treatise of Man* (pages 1 to 107)

The crux of Descartes's endeavor, in his *Man*, was the interpretation of physiological function in terms of matter in motion. Of *matter* he acknowledged three kinds, or elements, distinguished by the respective shapes, sizes, and motions of the particles they comprise (4).* As for *motion*, God distributed a certain quantity of

---

* The numbers in parentheses in this section refer to the correspondingly numbered notes to the translation.

it at the time of Creation; thereafter it was transferred from body to body without change in amount. Descartes asserted that bodies moving rectilineally continue to do so unless something happens to deflect them. He also developed a set of rules for the movement of particles in fluids (38).

In applying physics to physiology, Descartes placed special emphasis on the heat that every physiologist of the period attributed to the heart. He viewed man's body as a constellation of corpuscularly constituted, mechanically interacting parts (5, 6). The action of every part in the ultimate analysis is corpuscular motion (9–11), the particles moving either individually as in volatilization and condensation (22), fermentation (14, 19), and digestion (12, 13); or else en masse as in muscular contraction and the flow of the blood. The cardinal tenet in Descartes's physiology was its assumption that the body is so arranged (6) that every part can be activated by the transfer to it of motion that is ultimately derived from the heat of the heart. And the cardiac heat was itself, for Descartes, a form of motion, namely the agitation of blood particles engaged in active fermentation (25). So important seemed the heat — or, as he often called it, the "fire without light" — of the heart that Descartes sometimes equated it with life itself (20, 21, 26).

We may think of the cardiac cycle, in Cartesian physiology, as commencing when the ventricles are empty except for a small residuum of unevacuated blood. This residuum is important because it communicates its fermentive activity to the whole mass of blood that presently enters. The result, when the heart has been filled, is a general dilation of the blood-mass and of the heart itself (25, 30). The expanded and agitated blood escapes, and its transfer to the arteries makes the latter swell simultaneously throughout the body; this instantaneous arterial inflation is the pulse (31). Blood that moves from the right ventricle to the lungs is condensed there, and as its particles lose their agitation and heat, they aggregate as drops (22) which "fall" (23) into the left side of the heart (30).

In all regions of the body, certain particles escape from the small peripheral arteries and are added to the solids, fluids, and spirits (15, 33); for in living things all parts are continually displaced and replaced. Not only are the tissue fibers and the fluids formed in this way; so are such useful secretions as the juices aiding digestion (35,

36). In the stomach, digestive juice acts on the food in a manner similar to the action of acid on metals (12); this action is abetted by heat (13). Absorption from the stomach and intestine entails a differential diffusion similar to the sifting of matter through a sieve (16–18, 119). Blood formation occurs in the liver through fermentive conversion of chyle, although some chyle mixes with blood the moment it is absorbed (16, 19).

In general, the *Treatise of Man* is topically unbalanced. The discussion of digestion is pregnant, but brief. Breathing is treated at some length — especially in its muscular aspects — but Descartes was less fertile concerning its usefulness (120) than earlier authors, especially Galen. Transport within the intestine is covered in a sentence or two, as is the whole subject of excretion, despite the emphasis given it by Galen. Reproduction and embryogenesis are sidestepped. Descartes was to write about them later after further study made him surer of his ground (39).

The operations most extensively analyzed are those of muscles and nerves. Indeed, it would not be inaccurate to characterize our *Treatise* as an essay in physiological psychology with supportive — but subordinate — sections on other physiological topics. Descartes regarded every peripheral nerve as combining both sensory and motor functions (47) (although he did not use the terms peripheral, sensory, or motor). He viewed each nerve as a hollow tubule with a sleevelike double outer sheath, the inner and outer membranes of the nerve sheath being continuous with the inner and outer meninges of the brain. Each nerve tubule contains a central "marrow" of longitudinal fibrils surrounded and protected by animal spirits moving slowly outward from the brain (44, 46, 124).

The spirits are composed of highly volatile material particles derived from the blood (37, 43, 122). Some enter the ventricles from the minute arteries of the plexuses that carpet the ventricular cavity (128, 129). Others are derived from the arteries surrounding and supporting the pineal gland (which Descartes erroneously placed inside the cavity of the brain) (40–42); these spirits first enter the gland itself, then the ventricles (124–127, 130). From the ventricles, spirits move through the interfibrillar spaces of the brain substance into the channels of the cranial nerves and, by way of the spinal cord, those of the spinal nerves as well (44, 45). They keep the brain tense and turgid except during sleep when, because the flow is reduced, it becomes rather flaccid and relaxed (131, 156).

External motions (we should say stimuli) may displace the peripheral — and ipso facto the central — ends of the nerve fibrils. Centrally, the fibrils are arranged somewhat like broom whiskers (125) pointing inward; their central ends mark the outer boundaries of the ventricles of the brain (58, 60). The displacement of these central endings results in a differential pattern of openness of the interfibrillar spaces and hence in a preferential flow of spirits into some spaces, and into certain nerves, rather than into others (60, 61, 104, 124). Descartes's postulation of this schema, by which the peripheral disturbance of a particular nerve may lead automatically to a flow of spirits into the same or other nerves, has caused him to be called the founder of reflex theory. He personally never used the French noun *réflexe* in this connection. In his *Passions of the Soul*, however, he did speak of the pattern of events as being in some sense reversed or turned back (*réfléchis*) at the surface of the pineal gland in automatic responses (AT 11:356).

The soul, an entity separate from the body (2, 64) and resident in the gland (135), may or may not become aware of the unequal outflow of spirits; if it does, the result is conscious sensation (104, 133–136, 145). The soul in turn may initiate a differential outflow from the gland through the ventricles, and into the channels of particular nerves; this flow results in voluntary muscular action (44, 45, 124, 139–141, 143, 155). A muscle undergoing contraction is volumetrically inflated by taking up animal spirits. Only a small portion of these come to the muscle from its nerve. The bulk come, rather, from the muscle's antagonist by way of a short channel or shunt that Descartes imagined to exist between the two members of each antagonistic pair. An intricate valvular mechanism regulates the interflow of spirits between the two muscles; it is this mechanism that the spirits from the nerves control (48, 50–56, 130).

Descartes was especially interested in and willing to spell out the mechanisms of sensation and sensory discrimination (65, 72, 74, 147–149). He gives considerable detail on pain and pleasure (66), touch, temperature sensation (69), taste (74, 75), smell (76–82), and hearing — in connection with which he briefly outlines his theory of music (83, 85, 86). All these subjects are treated in the corpuscularizing manner he invariably adopts. The gross anatomy of the sense organs, and others, comes from Galen, Vesalius, and the post-Vesalian anatomists, although Descartes verified their findings, to an uncertain extent, by independent dissections of his own (see Foreword).

His principal concern was the eye. An elaborate and untenable corpuscular interpretation of light (87) serves as introduction to the functional anatomy of the eyeball (88–92, 94) and its muscles (93). Making use of the sine law of refraction, Descartes ably applied to the eye an improved adaptation of Keplerian optics. He gives praiseworthy but not always correct interpretations of image formation (96), optical definition (101, 146), depth and distance perception (99, 100, 106, 111), shape and size (102), position (107), perspective (103, 112), and, less convincingly, the transfer of the visual image to the brain (132).

Descartes retained a traditional distinction between external and internal sensation (113) — the latter including not only hunger and thirst (115) but also memory and emotion. He was later to write a treatise, the *Passions of the Soul*, analyzing these subjects in their conscious aspects (158). In *Man* he discusses their physiological basis (116, 151, 153, 154) and their dependence on the abundance and constitution of spirits (117, 118, 121, 122). He gives a suggestive model for the hypostasis of memory, wrong in its details but right in its assumption that a physical basis of retention must exist (137, 138). Near the end of the treatise he introduces a historically important analysis of instinct (150, 152).

(4) Text of the *Description of the Body* (pages 109 to 170)

In 1648, Descartes mentioned in a letter to the Princess Elizabeth that he was involved in a new work, a reconstruction of an earlier physiological treatise (*Man*) the manuscript of which, in partial disarray, had "fallen into the hands of several persons who transcribed it badly" (AT 5:112). The manuscript of the new work, the *Description of the Body*, came to France from Stockholm after the death of Descartes, in the same folder with the manuscript of *Man*. Clerselier decided to publish the two treatises together.

Comprising five parts, the *Description* begins (part I) with a defense of the anti-animistic assumptions that are central to the author's physiological point of view and a statement of the topics he intends to consider. Next follow two physiological sections (parts II and III). These are elaborations, respectively, of what Descartes had said in *Man* about circulation and nutrition. There is some evidence that the *fact* of circulation had occurred to Descartes before he knew of Harvey's proofs of it (Rothschuh, D., 27 and 48–49). Descartes claims no priority in the matter, however, and credits Harvey with having "broken the ice." However, he disagrees with Harvey concerning the physical causes of the heartbeat, the pulse, and the movement of the blood. He ascribes all these, as we shall see, to heat. In general, wherever the *Description of the Body* departs significantly from *Man* on the subjects of circulation and nutrition, we shall try to indicate the differences in our notes.

Remarkably, the very topics treated most extensively in *Man* (nervous action and brain function) are largely omitted from the *Description of the Body*. But in parts IV and V of the *Description*, Descartes presents at length his generally very implausible ideas on reproduction and the early stages of development, subjects he had steered clear of, for the most part, in *Man*.

He considered the chief initiative event in reproduction to be a mixing of male and female semina, each of which behaves as a ferment with respect to the other. The result of this fermentive encounter is a local redistribution of particles that leads to the formation of the heart. From the heart, currents move through the seminal *mélange*, laying out the future organs. These organs, initially of a fluid consistency, acquire stability and firmness through a dual process of enduration and fiber formation.

The general presentation of embryogenesis just outlined is not completely original with Descartes; nor would his ideas seem totally unwarranted in view of their date, were it not for his further speculations on specific formative events which are described in surprising and totally unverified detail. These speculations illustrate with tragic clarity the hopelessness of the author's epistemological or methodological goal. At the beginning of his intellectual career he was inspired — partly by a famous dream that he had while traveling abroad — to analyze the whole world and man by deducing their essential nature from a few self-evident axioms. His *Description of the Body* clearly proves that this goal was unattainable.

Variations in the title. In Clerselier's first French edition, this treatise begins on page 109 with the title *Description of the Human*

Body and All Its Functions; Both Those That Do Not, and Those That Do, Depend on the Soul. Also the Principal Cause of the Formation of Its Parts. On all subsequent pages, the treatise carries an entirely different running head, namely *Of the Formation of the Fetus.* In addition to being used as a running head, *De la formation du foetus* or its Latin or Dutch equivalent appears on the combined title pages of Clerselier's and all subsequent editions in which the two works were published together (see list of editions). Hence the treatise came to be known as *Of the Formation of the Fetus,* even though *Description of the Body* is more indicative of its total contents and is probably, most authorities agree, the title Descartes intended.

(5) *Remarks* of Louis de La Forge (pages 171 to 408)

The commentary of illustrator and annotator La Forge takes the form of footnotes that give an incomplete and rather unbalanced *explication de texte.* Some of the notes are elaborations on the mechanical explanations offered by Descartes. They purport to show how the sometimes sketchy Cartesian models of function can really be made to work. Some of the notes defend Descartes against his contemporary critics, among whom Bartholin is mentioned by name. La Forge treats the function of the brain at length, as does Descartes. Finally, certain of La Forge's *Remarques* depict and defend Descartes's explanatory method which, in his physiology as in his geometry, physics, and cosmology, begins with certain self-evident axioms and, using these as a basis, purports to build a theory of body function in general. In elaborating upon Cartesian hypotheses, La Forge exceeds his master in the invention of unverified detail. As to his illustrations of the text, La Forge explains that in making them he felt "committed less to representing things according to Nature than to rendering intelligible" what Descartes had to say (*l'Homme,* 326).

(6) *Preface* to Schuyl's Latin edition (pages 409 to 448). This appears in a French translation by Clerselier's son.

Schuyl's preface is an important resource for historians of physiology and physiological psychology. He focuses primarily on the soul-body problem, citing scriptural, patristic, and Scholastic evidence against the Greek idea that plants and animals have souls, at least of the conscious or cognitive sort that is present and operative in man.

The "souls" of animals, says Schuyl, are emergent manifestations of the material configuration of an immanently soulless body. As such, the souls of plants and animals are entirely distinct from the conscious souls with which God endowed human beings. The activities of plants and animals are attributable not to soul, but entirely to the organization of their bodies. By the same token, all those activities that man shares with plants and animals occur in man, too, without the intervention of soul. It was Descartes, says Schuyl, who first spelled out in corpuscular terms the latent basis of the patent acts of the body. Descartes thus saved for the soul that divine — and consciously cognitive role — that ecclesiastical dogma demanded.

BIBLIOGRAPHIC MATERIALS

EDITIONS OF THE *TREATISE OF MAN*

I. *Latin editions*

1662 *De homine figuris et latinitate donatus a Florentio Schuyl*, Leyden, Franciscum Moyardum et Petrum Leffen.

1662 The same. Another issue.

1664 The same, Leyden, *ex. off.* Hackiana.

1677 *Tractatus De homine et De formatione foetus: Quorum prior notis perpetuis Ludovici de La Forge, M.D. illustratur*, Amsterdam, Elzevir (a fresh translation).

1686 The same, Amsterdam, *ex. typ.* Blaviana.

1692 The same, Frankfurt, *sumpt.* F. Knochii.

1697 The same. Another issue, included in *Opera philosophica omnia*, Frankfurt, vol. 3, various pagings, with the title page of *Tractatus de Homine* bearing the date 1692.

II. *French editions*

WITH LA FORMATION DU FOETUS:

1664 *L'Homme de René Descartes: Et un traitté De la formation du foetus du mesme auteur: Avec les remarques de Louis de La Forge, docteur en médecine, demeurant à La Flèche, sur le traitté De l'homme de René Descartes et sur les figures par luy inventées*, Paris, Charles Angot.

1664   The same. Another issue, Paris, Nicholas Le Gras.

1664   The same. Another issue, Paris, Théodore Girard.

1680   *Les traitez De l'homme et De la formation du foetus, composez par M^r. Descartes, & mis au jour depuis sa mort par M^r. Clerselier: Avec les remarques de Louis de La Forge, docteur en médecine, sur le traité De l'homme du même auteur, & sur les figures par lui inventées*, Amsterdam, Guillaume le Jeune.

1729   *L'homme de René Descartes et La formation du foetus: Avec les remarques de Louis de La Forge. Nouvelle edition revûe et mise en meilleur ordre par Claude Clerselier*, Paris, Compagnie des Libraires.

WITH LA FORMATION DU FOETUS AND TRAITÉ DE LA LUMIÈRE:

1677   *L'homme de René Descartes et La formation du foetus: Avec les remarques de Louis de La Forge: A quoy l'on a ajouté Le monde ou Traité de la lumière du mesme autheur. Seconde édition, reveuë & corrigée*, Paris, Charles Angot.
       The same. Another issue, Paris, Théodore Girard.
       The same. Another issue, Paris, Michel Bobin & Nicolas Le Gras.

WITH MORE THAN TWO OTHER WORKS:

1824   *L'Homme* in Victor Cousin, ed., *Oeuvres de Descartes*, Paris, F. G. Levrault, vol. 4, pp. 335–428. Issued the same year in Strasbourg.

1909   *L'Homme* in Charles Ernest Adam and Paul Tannery, eds., *Ouevres de Descartes*, Paris, Leopold Cerf, vol. 11, pp. 119–202, with various tables, pp. 203–215.

1957   The same, Paris, Vrin.

1967   The same, Paris, Vrin.

1953   *L'Homme* in André Bridoux, ed., *Oeuvres et lettres*, Paris, Gallimard (Bibliothèque de la Pléiade), pp. 805–873.

1963   *L'Homme* in Ferdinand Alquié, ed., *Oeuvres philosophiques de Descartes: Texte établi, presenté et annoté par F. A.*, Paris, Garnier Frères, vol. 1, pp. 379–480.

1966   *L'Homme* in Samuel S. de Sacy, ed., *Oeuvres de Descartes: Edition établie, presentée et annotée par S. S. de S.: Avec quatre introductions par Geneviève Rodis-Lewis*, Paris, Club français du livre, vol. 1, pp. 345–401.

III.   Dutch editions

1682   *De verhandeling van den mensch en De makinge van de vrugt in's moeders lichaam . . . in't ligt gegeven door Clerzelier: Met de aan-*

teikeningen van Ludowyk de La Forge over het Tractaat van der mensch . . . in't Nederduits overgezet door Jacob Copper, Middelburg, [widow of] R. Schrijver & A. Rammazein.

1695 The same. Another edition, Leyden, Frederik Haaring.

1690– *Konstig geboux des menschelijken lighaams* . . . , trans. J. H.
1692 Glasemaker, in *Alle de werken van de Heer Renates Des-Cartes,* Amsterdam, Jan ter Hoorn, vol. 3, pp. 221–296.

IV. German edition

1969 *Über den Menschen (1632), sowie Beschreibung des menschlichen Körpers (1648): Nach der ersten französischen Ausgabe von 1664 übersetzt und mit einer historischen Einleitung und Anmerkungen versehen, von Karl E. Rothschuh,* Heidelberg, Lambert-Schneider.

V. Italian edition

1960 *L'Uomo: Introduzione e traduzione di Gianfranco Cantelli,* Turin, Boringhieri.

VI. English edition

1972 *Treatise of Man: French Text with Translation and Commentary by Thomas Steele Hall,* Cambridge, Mass., Harvard University Press.

SOURCES FOR THE STUDY
OF DESCARTES'S PHYSIOLOGY

Additional physiological writings of Descartes:

The *Discourse on Method,* section V, 1637 (summarizes portions of *Man*).

The *Dioptrics,* 1637 (but written before 1630 — that is, before *Man* — one of the three works to which the *Discourse* was published as an introduction; treats extensively the eye and the general subject of sensation).

The *Principles of Philosophy,* 1644 (gives in capsule form important elements of Descartes's theory of physiological psychology).

The *Passions of the Soul,* 1649 (partially completed 1645–1646; contains sections of physiological psychology).

Certain notes made at various times and not intended for publication (see *Anatomical Excerpts* in the Abbreviations of Reference Works).

The *Description of the Human Body*, composed about 1648 (first published with *Man* in 1664).

Descartes's letters (many of which have to do, in whole or in part, with physiological questions).

There is also a fragment, *First Thoughts on the Generation of Animals*, probably composed as early as 1629. Many of the ideas contained therein were reconsidered later by the author and did not become a part of his mature physiological theory. Not long before his departure for Sweden, Descartes composed a set of notes and commentaries on the mind-body problem which were published in Amsterdam in 1648 under the title *Notae in programma quoddam sub finem anni 1647.* . . .

Selected list of secondary sources:

Ayrer, A. F., *De quibusdam physiologiae Cartesianae capitibus dissertatio* . . . , Göttingen, J. C. Dieterich, 1791.

Baillet, A., *La vie de M. Des-Cartes*, Paris, D. Horthemels, 1691.

Berger, A. M. J. von, *Hielt Descartes die Theorie für die Thiere für bewusstlos?*, Vienna, Kaiserliche Akademie den Wissenschaften, Sitzungsberichte, Bd. 126, 1892.

Beverwyck, J. van, *Epistolicae quaestiones cum doctorum responsis* . . . , Rotterdam, A. Leers, 1644.

Böhm, W., "John Mayow und Descartes," *Sudhoffs Archiv*, 46(1962), 45–68.

Borel, P., *Historiarum et observationum medicophysicarum* . . . *centuriae IV* . . . , Paris, M. Dupuis, 1656.

Chauvois, L., *Descartes: Sa méthode et ses erreurs en physiologie*, Paris, Éditions du Cèdre, 1966.

Condillac, Étienne Bonnot de, *Traité des animaux* . . . , Amsterdam and Paris, G. F. de Bure, 1755.

Cordemoy, G. de, *Le discernement du corps et de l'âme* . . . , Paris, F. Lambert, 1666.

Crombie, A. C., "Descartes," *Scientific American*, 201(1959), 160–173.

—— "The mechanistic hypothesis and the scientific study of vision . . . ," in S. Bradbury and G. L'E. Turner, eds., *Historical Aspects of Microscopy*, Cambridge, W. Heffer for the Royal Microscopical Society, 1967, 66–112.

—— "Some aspects of Descartes' attitude to hypothesis and experiment," *Collection des travaux de l'Académie internationale d'histoire des sciences*, Florence, Bruschi, 1960, 192–201.

Daniel, G., *Voyage du monde de Descartes*, Paris, [veuve de] S. Bernard, 1690, trans. T. Taylor, *Voyage to the World of Descartes*, London, T. Bennet, 1692.

Dankmeijer, J., "Les travaux biologiques de René Descartes," *Archives internationales d'histoire des sciences*, 16(1951), 675–680.

Dreyfus-Le Foyer, H., "Les conceptions médicales de Descartes," *Revue de métaphysique et morale*, 44(1937), 237–286.

Georges-Berthier, A., "Le mécanisme cartésien et la physiologie au 17ème siècle," *Isis*, 2(1914), 37–89 and 3(1920), 21–58.

Gilson, É. H., *Étude sur le role de la pensée médiévale dans la formation du système cartésien*, Paris, Vrin, 1930 and 1951.

———— *Index scholastico-cartésien*, New York, Franklin, 1964.

———— footnotes to pt. 5 of his *René Descartes: Discours de la méthode*, Paris, Librairie Philosophique, 1930.

Haldane, E. S., "Descartes' physiology," chap. 16 in *Descartes: His Life and Times*, London, Murray, 1905.

Haldane, J. S., "The physiology of Descartes and its modern developments," *Acta biotheoretica*, 1(1935), 5–16.

Hall, T. S., "Microbiomechanics," chap. 4 in *Ideas of Life and Matter*, Chicago, University of Chicago Press, 1969, 250–263.

———— "Descartes' physiological method . . . ," *Journal of the History of Biology*, 3(1970), 53–79.

Jaumes, A., *De l'influence des doctrines philosophiques de Descartes et de Bacon sur les progrès de la médecine*, Montpellier (thesis), 1950.

Jens, P., *Enchiridion philosophicum seu aphorisimi . . .* , Leyden, Haaring, 1690.

Jörges, R., *Die Lehre von dem Empfindung bei Descartes*, Düsseldorf, L. Schwann, 1901.

La Forge, L. de, *Traitté de l'esprit de l'homme . . .* , Paris, M. Bobin and N. Le Gras, 1661.

Liesegang, G., *Descartes' Dioptrik*, Meisenheim an Glan, 1954.

Mesnard, P., "L'Esprit de la physiologie cartésienne," *Archives de philosophie*, 13(1937), 181–220.

Pitcarnius, A., *Compendiaria et perfacilis physiologiae idea . . . una cum anatome Cartesianismi . . .* , London, T. Parkhurst, 1676.

Rosenfield, L. C., *From Beast-Machine to Man-Machine . . .* , New York, Oxford University Press, 1941.

Rothschuh, K. E., "René Descartes und die Theorie de Lebenserscheinungen," *Sudhoffs Archiv*, 50(1966), 25–42.

Saint-Germain, B. de, "*Descartes consideré comme physiologiste et comme médecin*, Paris, Masson, 1869.

Sánchez Vega, M., "Estudio comparativo de la concepción mecanica del animo y sus fundamentos en Gomez Pereyra y Renato Descartes," *Revista de filosofía*, 13(1954), 359–508.

Schoock, M., *Admiranda methodus novae philosophiae Renati Des Cartes*, Utrecht, J. van Waesberge, 1643.

Scott, J. F. *The Scientific Work of René Descartes, 1596–1650*, London, Taylor and Francis, 1952.

Sebba, G., *Bibliographia Cartesiana: A Critical Guide to the Descartes Literature . . .* , The Hague, Nijhoff, 1964.

Vartanian, A., *Diderot and Descartes: A Study of Scientific Naturalism in the Enlightenment*, Princeton, N.J., Princeton University Press, 1953.

TREATISE OF MAN
René Descartes

MAN

by

René Descartes

These men[1] will be composed, as we are, of a soul and a body;[2] and I must first separately describe for you the body; then, also separately, the soul; and finally I must show you how these two natures would have to be joined and united to constitute men resembling us. [Cl–1; AT–119]

NOTE: The bracketed references in the translated text are to page numbers in Clerselier's first French edition of *l'Homme* (designated "Cl–" and reproduced at the back of this book) and in Adam and Tannery's edition in the eleventh volume of the *Oeuvres* (designated "AT–").

1. *Descartes on the nature of man.* Descartes distinguishes between actual men (whom he often refers to as "us") and conceptual analogues or models thereof. He also contrasts his own Cartesian model ("these men," who will form the subject of his book) with the conventional model (men as ordinarily conceived by philosophers and physicians). He follows a parallel practice in the *Treatise of Light*, composed for copublication with *Man*, contrasting there the actual world with the Cartesian and conventional models or concepts of that world. The central question then is, which model — the Cartesian or the conventional — is more precisely analogous to the actuality?

Although he is hesitant to say so, Descartes wants the reader to accept the Cartesian alternative rather than the conventional view of man and the world. In the *Discourse on Method* (1637), which contains paraphrases of parts of *Light* and of *Man* (1629 to 1632), he gives his reasons for writing in this tentative and analogical fashion. Partly he wishes to shield himself, and especially his ideas, against ecclesiastical opposition (AT 6:60 ff.). Partly he wishes to acknowledge that all he can offer, epistemologically, is a conceptual parallel or metaphor of reality; he does not pretend to portray reality itself (AT 6:45; see also *Principles*, AT 8:327 [9:322–323]).

Descartes's commencement of this chapter with the demonstrative "These" is interpreted by Cartesian scholars as a back-reference to an earlier discussion of hypothetical analogues of real men, presumably in the missing two chapters by which the

1

I assume their body[3] to be but a statue, an earthen[4] machine[5]

Treatise of Man was connected with the *Treatise of Light* (AT 11:III–IV and XII; also Rothschuh, D., 43, note 1).

2. *Ideas of body and soul.* Few ideas of Descartes were to influence human thought more decisively and lastingly than his ideas about the body and the soul, their ontological separateness, and their modes of interaction. The somatopsychic dualism of Descartes should be viewed in the light of contemporary theories of "soul." His own view represents a response to an initially Greek but persistent tradition which regarded soul as both (*a*) the motive cause of physiological function and (*b*) the conscious agent of perception, volition, and reason. The crux of Descartes's endeavor is to eliminate the physiological role altogether and to limit the cognitive role to man (see his *Response to the 5th Objection to His 2nd Meditation*, AT 7:250–261; also Hall, 1970). Descartes usually portrays plants and animals as soulless automata (see *Discourse*, AT 6:46; *Response to the 4th Objection to His Meditations*, AT 7:229–230; *Passions*, AT 11:330–334; and letter to More, AT 4:573), although he sometimes acknowledges a corporeal soul in animals that distinguishes them from man, whose soul is incorporeal (see Schuyl in his preface to the Latin edition of *Man*). In a letter to Regius (AT 3:370) we hear that the powers of vegetation and of feeling in animals do not merit the designation "soul" in the sense that *mind* merits that designation in man.

From Schuyl, Descartes's editor and Latin translator, we learn that the Greek belief in a partial identity of beast soul and human soul had long been under discussion and debate. As early as the fourth century A.D., St. John Chrysostom had attributed this idea to the devil, who had managed, according to St. John, to deceive Zoroaster, Pliny, Pythagoras, Plato, Plutarch, Porphyry, Lipsius, and the Peripatetics on the matter (*de Homine*, 1662, fourth [unnumbered] page of Schuyl's preface "Ad Lectorem." This passage appears in translation in the first French edition of *l'Homme* on pp. 412–413; see also pp. 418–419).

In chapters 26 to 30 of his *De quantitate animae*, St. Augustine (d. A.D. 430) had emphasized the distinctive difference between mere sensation (as it occurs in animals) and reason (as in man), but he did not deny that animals have a soul. In 1554, Gomez Pereira published an explicit objection to the existence of a sensitive soul in animals (see the "Paraphrasis in tertium librum de anima Aristotelis longe ab omnium aliorum authorum expositione dissidens," in his *Antoniana margarita . . .*, Medina del Campo, 1554, columns 497–574); but in 1641, Descartes said, "I have not seen the *Antoniana margarita*, nor do I believe I have a great need of seeing them" (letter to Mersenne, June 23, 1641, AT 3:386).

For Descartes elsewhere on the somatopsychic duality of man, see his letter to *** March 1638, AT 2:37–42. See also, on the soul-body problem, Rothschuh, D., 43–44, note 2. Descartes had intended to extend *l'Homme*, which deals not only with physiology but also in a pioneering way with physiological psychology, to include a treatment of the soul in itself (see note 158). The intent bore fruit only later in his *Passions de l'âme*.

3. Not man's body, but that of the hypothetical analogue discussed in note 1.

4. *Theory of matter.* Descartes assumes three elements differing in the *size* and *shape* (and resultant *movement* or lack of movement) of their constituent particles. In his earlier works, he called these elements — in ascending order of particle size — Fire, Air, and Earth. He classified bodies in general as "pure" or "mixed."

(*a*) "Pure" (or uniform) bodies could consist of: just *one* element (the sun and fixed stars consist of first-element or Fire particles only); or *two* elements (the separate "heavens" revolving around the sun and around each fixed star consist primarily of second-element or Air particles, but with first-element particles filling up the interstices between them); or *three* elements (the earth, planets, and comets consist primarily of third-element particles, presumably with second-element particles partially filling the interstices, the rest of the space being occupied by particles of Fire).

(*b*) "Mixed" bodies exist at the interface between the earth and the heavens and comprise the living and nonliving objects that we see everywhere about us.

The ultimate result of this arrangement is that all space is filled. Descartes denied the existence of empty space anywhere.

Of the first element, Descartes advises us that "in order not to be forced to admit the existence of a void, I do not attribute [to this element] particles of determinate size or shape, but am persuaded that the impetuosity of its movement is sufficient to cause it to divide in every manner and every direction on encounter with other sorts of bodies; and [I am persuaded] that the parts change their shape from moment to moment in order to accommodate themselves to the spaces they enter; whence there is never a passage so narrow or angle so acute betwixt the parts of other bodies but that the parts of this element enter easily and fill these spaces exactly" (*Light*, AT 11:24; see also *Discourse*, AT 6:43; and *Meteors*, AT 6:233 ff.).

Second-element particles are of determinate shape tending to roundness. These particles are highly and equally responsive to forces that tend to (*a*) augment their movement and diminish their size, or (*b*) diminish their movement and augment their size. The effect of these opposing influences is to maintain the particles in a condition of intermediacy, with respect to size and movement, between the particles of the other two elements.

Third-element (Earth) particles are joined in large aggregates that, in contradistinction to second-element aggregates, are resistant to the movements of other, impinging bodies.

Lastly, Descartes makes the point that elementary Fire, Air, and Earth are not the fire, air, and earth with which our daily experience makes us familiar; the latter "are mixed and composite and are subject to corruption." This is not to imply that the three elements are nontransmutable. Under proper conditions, Earth particles are reducible to those of Air, and those of Air to those of Fire.

When Descartes later returned to his theory of elements (in the *Principles*), he kept the main essentials but changed his terminology somewhat. In order to avoid confusion with the familiar connotations of "fire," "air," and "earth" he called his three sorts of matter the "first element," the "second element" (sometimes *matière du ciel*), and the "third element" (sometimes *matière terrestre*); see *Principles*, AT 8:100–105 (9:124–129). As to the *tria prima* of the chemists and the *four elements* of the philosophers, he believed that there was ultimately but one sort of matter, of which various kinds differed only in the shapes and arrangements of their constituent particles (letter to Newcastle, AT 4:569–570, and to Mersenne July 30, 1640, AT 3:131).

For more on Descartes's theory of matter, see M. Boas, "The establishment of the mechanical philosophy," *Osiris*, 10(Bruges, 1952), 422 ff.; J. R. Partington, "The origin of the atomic theory," *Annals of Science*, 4(1939), 245; E.J. Dijksterhuis, *The Mechanization of the World Picture* (Oxford, 1961), 409 ff.; and R. H. Kargon, *Atomism in England from Hariot to Newton* (Oxford, 1966), 64–65.

5. *Automata*. Mechanical analogues of men and animals were familiar to Descartes, especially in articulated clock- and garden-figures that produced the illusion of self-instigated movement (see also notes 2 and 7). Such figures, variously moved, belonged to many cultures and had long histories dating back to primitive, usually religious, articulated statues and figurines. Schuyl mentions a number of descriptions of automata, some of which Descartes may have been familiar with (*de Homine*, 1662, commencing with the tenth [unnumbered] page of the "Ad Lectorem").

For a history of automata, see A. Chapuis and É. Gélis, *Le monde des automates*

formed intentionally by God to be as much as possible like us. Thus not only does He give it externally the shapes and colors of all the parts of our bodies; He also places inside it all the pieces required to make it walk, eat, breathe, and imitate whichever of our own functions can be imagined to proceed from mere matter and to depend entirely on the arrangement of our organs.[6] [Cl-2; AT-120]

We see clocks, artificial fountains,[7] mills, and similar machines which, though made entirely by man, lack not the power to move, of themselves, in various ways. And I think you will agree that the present machine could have even more sorts of movements than I have imagined and more ingenuity than I have assigned, for our supposition is that it was created by God.[8]

Now I shall not pause to describe to you the bones, nerves, muscles, veins, arteries, stomach, liver, spleen, heart, brain, nor all the other different pieces of which the machine must be composed; for I suppose them all to be quite like the parts of our own body that have the same names. If you do not already know them sufficiently, you can have them shown to you by some learned anatomist, those at least that are large enough to be seen. As for those which because of their smallness are invisible, I shall be able to make them known to you most simply and clearly by speaking of the movements which depend upon them; so that it remains only for me to explain these

(Paris, 1928); also the article "Automata," *Encyclopedia of World Art*, vol. 2 (New York, 1960), 182–193. Mechanical models as devices in physiological interpretation go back at least to Aristotle ("The movements of animals may be compared with those of automatic puppets . . . or with the toy wagon." Farquharson trans., *MA* 701b1-4). For a modern appraisal of seventeenth-century automatism and mechanicism, see D. J. de S. Price, "Automata and the origins of mechanism and the mechanistic philosophy," *Technology and Culture*, 5(1964), 9–42; see also note 152 below.

6. Descartes here reinitiates an old debate as to whether organization produces life or vice versa. The issue had its roots in Greek atomism and received its most explicit statement, in antiquity, in the world-scheme of Epicurus (see esp. Lucretius, *De natura rerum*, ii, 865–885 and iv, 822–857). Louis de La Forge, the disciple of Descartes who annotated his *Man*, defines a machine as any "body composed of several organic parts which being united conspire to produce certain movements of which they would be incapable if separate" (*l'Homme*, 132).

7. Descartes was especially impressed with the fountains at *Saint-Germain-en-Laye*, which he knew either personally or from the work of Salomon de Caus (see frontispiece).

8. For more on the complexity and ingenuity of the machine see the *Discourse*, AT 6:56.

movements to you here in proper order and by that means to tell you which of the machine's [latent] functions these [patent] movements represent.[9]

First, then, the food is digested in the stomach of this machine by the force of certain liquids[10] which, gliding among the food particles,[11] separate, shake, and heat them just as common water

9. *Visible vs subvisible.* Descartes here reveals his analytical program: *he will account for the visible actions of visible organs in terms of the invisible actions of structures too small to be seen.* He devotes several articles in his *Principles* to an explanation of this procedure as he follows it in both his physiology and physics (AT 8:325–326 [9:321–322]).

La Forge, in his commentary published with the first edition of *Man*, explains and defends Descartes's postulation of subvisible body mechanics, and sets up four criteria by which conceptual models of such mechanics should be measured: he says that (*a*) they should be mechanically operable, (*b*) they should not contradict what gross anatomy reveals, (*c*) they should accurately and easily explain the things they are supposed to explain, and (*d*) they should do so with greater ease and precision than any other discoverable explanations (*l'Homme*, 214–218, 308, and 406).

For discussions of Descartes's explanatory method, see Crombie, 1960; Hall, 1970; and Rothschuh, D., 44, note 1, and 45, notes 1 and 2.

Descartes stood at the threshold of an era in which it would be possible to seek optical evidence for some, but by no means all, of the things he set out to describe. As for Descartes, the subvisible world that was so basic to his whole endeavor and that he portrayed in such uninhibited detail remained a world of inference and imagination. This is true even though he was interested in and made some use of magnifying glasses. Thus, in the *Dioptrics*, he proposed a single-lens lunette for enlarging small objects held very close to the eye, the anterior lens surface being a concave hyperbole with the object at the focal point while the posterior surface (the one toward the eye) was supposed to be flat (AT 6:199–201). As to the benefits Descartes expected from the use of magnifying glasses, see AT 6:226.

10. *Mechanics and chemistry.* A tendency was to develop among post-Cartesian physiologists to choose between iatromechanical and iatrochemical interpretations of biological and medical phenomena. But there was also a tendency, on the part of some, to bring the two viewpoints together — specifically on the level of subvisible organization, where mechanics and chemistry appeared to many sevententh-century authors to be one and the same. For example, Descartes proposes here a mechanical (corpuscular) interpretation of phenomena at that time usually regarded as chemical. For another example, see the *Principles* (AT 8:256–258 [9:250–252]).

11. *Particle theory.* Cartesian physics places special emphasis on parts (*parties*). A part is "whatever is joined together," that is, any body not tending to undergo division (hence tending to move as a unit). In the visible realm, parts vary upward from motes in air — or, at a higher level, granules of sand — all the way to the earth as a whole. In the subvisible realm, they vary downward, the smallest being the constitutive particles of Descartes's three elements (three sorts of matter).

While Descartes was a corpuscularist, he was not an atomist, since he set no lower limit of divisibility on the corpuscles whose existence he assumed. Atomism in the strict sense, introduced into Greek thought by Leucippus and Democritus and later elaborated by Epicurus, had been reformulated in various ways and reintroduced to Western thought by persons slightly antecedent to or contemporary with Descartes, among them Francis Bacon, Galileo, Daniel Sennert, and Sebastian Basso. Descartes

does the particles of quicklime, or aqua fortis those of metals.[12] To which must be added the fact that these liquids, being brought quickly from the heart by the arteries, as I shall tell you hereafter, are necessarily very hot.[13] And the food is ordinarily of such a nature

learned about these authors and their ideas in some cases by reading their works and in others through his friend and mentor Isaac Beeckman and his correspondent Marin Mersenne. On this subject see the references to secondary sources contained in note 4.

Descartes extended his dependence on corpuscles (*particules*, *petites parties*) as explanatory devices to every branch of his science — cosmology, physics, chemistry, physiology. His evocation of particles appears on page after page of his works on these subjects. Yet he did not view particles as having set limits of divisibility; he explicitly opposed atomicity in this sense. For many references to this topic in Descartes's works, see Gilson, *ISC*, 34.

Historians have used the terms "corpuscle," "corpuscular," and "corpuscularism" in their discussions of particle theories beginning with those of ancient Greece. We shall follow their example in referring to Cartesian physics as corpuscular, even though to do so is in a sense anachronistic, the term having been given general currency (mainly through the influence of Boyle) only after Descartes's death.

12. Belief in the digestive effect of acidity was not original with Descartes, but his sources are difficult to trace with precision. Galen thought the spleen sent to the stomach a residue of black bile whose sharpness and sourness made the stomach contract and retain the food undergoing concoction. Although he assigned a certain fermentive role to acids, he thought that concoction itself was caused by heat (*UP*, bk. 5, chap. 4, K3, 361–363; *de Simp. med.*, K11, 453). A more definitely digestive role was assigned to acid substances in medieval thought. Such thinkers as Paracelsus and Fernel supposed that it was through the action of acids that animals were able to handle recalcitrant substances including — in the case of the ostrich — even iron (see note 13).

Nitric acid (aqua fortis) was familiar in Renaissance alchemy and metallurgy, especially as a solvent of silver. Descartes's contemporary van Helmont interpreted the acidity of the stomach as necessary to the action of an energizing ferment brought there from the spleen ("Sextuplex digestio alimenti humani," *Ortus medicinae*, Amsterdam, posth., 1648), but van Helmont's influence on Descartes is problematical. In effect, although we cannot be sure about Descartes's exact sources on acid digestion, we can say that he borrowed the idea and restructured it to make it fit his own assumptions about the corpuscular constitution of the body.

13. *Digestion and heat.* Various Greek thinkers had linked heat with digestion in various ways, each theory reflecting its author's ideas about the nature of matter in general. Thus Plato thought that sharp-edged (tetrahedral) particles of the fiery element chop up the food in the stomach and drive it into the body (*Timaeus*, 78e and 80d). Aristotle saw the gastric alteration of food as a "cooking" (pepsis) made possible by the body's innate heat, this coction being the first in a series that prepares the food for incorporation into tissue (*PA*, 350a5, 351b25). A different interpretation was offered by the late-Hippocratic author Diocles (circa 350 B.C.), who thought that heat was generated by the digestive action of absorbed pneuma on the food in the stomach (Clifford Allbutt, *Greek Medicine in Rome*, London, 1921, 239).

Galen followed Aristotle in viewing coction as an assimilation (a "like-making") with the difference that, according to Galen, the stomach (*a*) turns part of the food into something similar to stomach tissue itself, (*b*) incorporates that part, and (*c*) expels the rest into the intestine. The gastric coction is permitted, he thought, by a resident alterative faculty of the stomach. Concerning the assimilative process, Galen said that "just as fire makes wood like itself, taking from the wood both its

that it can be broken down and heated quite of itself just as new hay is if shut up in the barn before it is dry.[14] [Cl–3; AT–121]

Know too that the agitation that the food particles receive in being heated, together with the agitation of the stomach and bowels that contain them, and the arrangement of the fibers[15] of which the bowels are composed, cause these particles — in the

beginning and its nourishment, in the same way both plants and animals are observed to assimilate their food to themselves" (*de Plac.*, trans. De Lacy, bk. 6, chap. 6; Galen seems here to be paraphrasing Aristotle, *GC*, 322a10 ff.).

Among Descartes's more recent forerunners, Paracelsus viewed the stomach's action as a heat-induced separation of poisonous from nonpoisonous parts of the food, followed by rejection of the former part, and conversion of the latter into something not only nonpoisonous but useful. The responsible agent, he said, is "an alchemist" using the stomach as a workshop in which "he labors and boils" (*Volumen medicinae paramirum*, trans. K. F. Leidecker, Baltimore, 1949, 26–27).

Fernel (circa 1540) postulated a heat-supported "concoctive faculty," which (more or less following Galen) he thought of as a differentiation of soul. He added that this faculty operates in the stomach as in a cauldron (*Physiologie*, 518–519). Fernel also acknowledged an "occult property" of the stomach capable, like Paracelsus' alchemist, of handling poisonous and refractory ingredients of the food; it also enables the ostrich to digest iron, small birds to digest seeds, and the like (*Physiologie*, 519, and *De abditis causis rerum*, first publ. Paris 1548, 1560, 210).

The metaphors developed by Paracelsus and Fernel may or may not have affected Descartes directly. But they illustrate the way in which stubborn ideas such as the concept that heat and digestion are related, were developed by differently oriented pre-Cartesian inquirers. These thinkers adapted — we might almost say twisted — the heat-digestion idea to make it fit the general assumptions within which each of them reasoned.

The same thing may be said of Descartes, who continues — but alters — the established tradition on this subject. He envisions two sources of the heat: namely, that generated by spontaneous decomposition of the food itself and that communicated by the heart to certain juices in the blood which communicate it in turn to the food in the stomach. In both cases, the heat in question is an expression of particle motion, as indeed are all the functions of the body in his physiological scheme.

14. The heating of wet hay and other thermogenic reactions are attributed by Descartes to particle motion. For a full analysis, based on his particle theory, see the *Principles*, AT 8:256 (9:250–251). Earlier, Descartes had emphasized the indispensability for thermogenesis of a reaction between different — or even opposite — substances (see, for instance, *Anatomical Excerpts*, AT 11:631–632). For example, he believed for a time that the heart was formed in the embryo by a thermogenic interaction of two different substances derived from the previously formed lungs (vital spirit) and liver (blood) (*Generation*, AT 11:508–511 and 599). Later he considered it possible for almost identical substances such as the male and female seminal fluids to engage in mutual fermentation when mixed, and he decided that such a thermogenic mixture formed the heart, or at least the left ventricle, before any other organ was formed (*Description of the Body*, AT 11:252–254). See also notes 21, 25, and 26 below.

15. *Fibers.* Descartes assumes a fibrous structure for the solids, or at least for some of them. This idea was one with an already well-developed history. Galen, in an attempted reformation of even earlier ideas about fibers, had given muscles a fibrous structure at the subvisible level, supposing that within the muscle the sensitive terminal subdivisons of nerves combine with the insensitive terminal subdivi-

measure that they are digested — to descend little by little toward
the conduit through which the coarsest of them must go out [the
rectum]. [Know] too that the subtlest and most agitated particles
meanwhile encounter here and there an infinity of little holes
through which they flow into the branches of a large vein that carries them toward the liver, and into others that carry them elsewhere — nothing but the smallness of the holes serving to separate
these from the coarser particles;[16] just as, when one shakes meal in
a sack, all the purest part runs out, and only the smallness of the
holes through which it passes prevents the bran from following
after.[17]

---

sions of ligaments to form fibers that emerge from the farther end of the muscle as
tendons (de Plac., bk. 1, chap. 9, K5, 204). He then went further and, according
to Siegel (Galen's System, 234), considered the solid body parts to be composed of
membranes enclosing fibrils and interfibrillar flesh.

With the elaboration of Galen's doctrines in Renaissance Europe, variants of the
fiber theory were proposed by many anatomists including Fernel, Vesalius, Paré, and
especially Riolan. Indeed, the notion of fibers as fundamental structural elements
was to survive until the nineteenth century, when it was effectively dispensed with
by Schwann who argued that fibers were modified cells or parts or constellations of
cells. For important aspects of the history of fiber theory, see A. Berg, "Die Lehre
von der Faser als Form- und Funktions-Element der Organismus," Virchow's Archiv
für pathologische Anatomie und Physiologie, 309 (1942), 394 ff.

16. Absorption and sanguification. On this subject Descartes departs substantially from the opinions of his contemporaries. To understand the context within
which he developed his ideas, we need to remember that Galen had believed that
concocted nutriment was attracted to the liver and there converted to blood and the
three other humors (UP, bk. 4, chaps. 3, 12, and 13, K3, 269–270 and 296–311;
and bk. 5, chap. 4, K3, 351–362). Galen's ideas on this subject were rather generally
subscribed to, though with individual modifications, by a majority of pre-Cartesian
Renaissance medical theorists.

Rethinking the problem in the light of his own corpuscular theories, Descartes sees
chyle as mixed with blood during the blood's circulation through the intestinal walls.
Harvey held similar views about absorption — without, however, invoking corpuscles
or the idea of physical filtration (Motion of the Heart, chap. 16). Neither Harvey
nor Descartes, however, was ready to deny the liver a role in blood-making (for details see note 19). Before Descartes began writing his Man, Aselli had published his
discovery of the lacteals (De lactibus, sive lactis venis . . ., Milan, 1627), but he
mistakenly believed that they brought chyle to the liver. In any case, Descartes learned
of the lacteals only later, in 1640 (Georges-Berthier, 1914, note 3). In a letter to
Mersenne on July 30 of that year (AT 3:140–141), Descartes claims that both the
mesaraic veins and the lacteals absorb chyle, the difference being that in the veins
— but not in the lacteals — the chyle is mixed with blood. Pecquet's discovery of
the thoracic ducts (Experimenta nova . . ., Paris, 1651) and the working out of
the lymphatic system by Bartholin (De lactis thoracicis, Copenhagen, 1652; Paris,
1653) and Rudbeck (Nova exercitatio . . ., Uppsala, 1653) came too late to have
influenced Descartes in this early work.

17. In Descartes's letter to Mersenne of July 30, 1640, he ascribes what we call
absorption (a) to the agitation of the particles, (b) to their weight (as when weight

These subtler parts of the food, being unequal and still imperfectly mingled, compose a liquid which would remain quite turbulent and whitish were it not that a part of it is blended directly with the mass of blood that is contained [a] in all the branches of the portal vein (which receives this liquid from the intestines), [b] in the vein that is designated "caval" (this vein conducts it toward the heart), and [c] in the liver itself as if in a single vessel.[18] [Cl-4; AT–123]

Similarly, it should here be noted that the pores of the liver are so arranged that this liquid, when it enters, is subtilized and elaborated, and takes on the color, and acquires the form, of blood, just as the white juice of black grapes is converted into light red wine when one lets it ferment on the vine stock.[19]

Now the blood, thus contained within the veins, has but one manifest passage by which it may leave them, namely that which conducts it to the right cavity of the heart. Know too that the flesh of the heart contains in its pores one of those fires without light[20] of

---

pulls milk down through the walls of a porous vessel), and (c) to intra-abdominal pressure (AT 3:140–142). In using this sieve model of absorption (and excretion), Descartes notes that not only the size of the pores but also their shape may decide which particles pass through.

18. In a considerably later letter to Regius (May 24, 1640, AT 3:66–68), Descartes says that food becomes, first, *chyle* (this occurs in the stomach); then, after admixture with blood, *chyme* (this occurs in the liver through a kind of fermentation); and finally, *blood* (in the heart through an ebullient reaction). The comparison of the liver to a single vessel is surprising in that Galen had argued at length that blood moves through the liver by way of finely subdivided blood vessels and not as if passing into and out of a single vessel (*UP*, bk. 4, chap. 13, K3, 303–307).

19. Again, Descartes offers a corpuscular interpretation of chemical change (see also note 10). His comparison of blood formation with vinous fermentation is reminiscent of Galen, according to whom the chyle is resolved, in the liver, into three components comparable (a) to wine (blood), (b) to the foam or flower of the wine (yellow bile), and (c) to the leas (black bile) (*UP*, bk. 4, chap. 3, K3, 269–270). This comparison of the liver to a wine vat reappeared in neo-Galenic thought; see, for example, Fernel, *Physiologie*, 540–541. Van Helmont, a contemporary of Descartes, viewed hematopoiesis (blood formation) as one of six ascending fermentations (*Sextuplex digestio alimenti humani, Ortus medicinae*, Amsterdam, 1648). Descartes's indebtedness to Fernel has been thoroughly established; the extent of van Helmont's influence is uncertain.

20. *Fire.* In the *Principles* (AT 8:250 [9:244–245]), Descartes says that visible fire (flame) involves the continuous direct agitative action of first-element particles on third-element particles. For this to happen, it is necessary that particles of the second element be driven out, and kept out, of the body concerned. This can only occur, he believes, if the particles of the third element are large enough. Otherwise evaporation rather than combustion occurs, because escaping air particles carry the terrestrial particles along with them. (See also *Light*, AT 11:7–9.) Invisible fires,

which I have spoken to you heretofore which renders it so fiery and hot[21] that, in the measure that the blood enters either of the two chambers or cavities that are there, it is promptly inflated and dilated. In like manner, you can prove experimentally that the blood or milk of some animal will be dilated if you pour it drop by drop into a very hot flask. And the fire in the heart of this machine that I am describing to you serves no other purpose than to dilate, warm, and subtilize the blood that falls continually drop by drop through a passage from the vena cava, into the right cavity whence it is exhaled into the lung, and from the vein of the lung (to which anatomists have given the name "venous artery") into the other cavity whence it is distributed through the body. [Cl–5; AT–123]

---

like visible ones, entail a direct action of first-element particles on third-element particles. See also note 14.

21. *The heat of the heart.* According to modern physiology, the blood is not changed by its passage through the heart, except for variations in the pressure exerted on it. But the realization that this is so required a reversal of traditional views according to which the blood in the heart was both heated and transformed. As to heat, Aristotle said that "during life animals are warm, but when dead . . . cold. The source of this heat in sanguineous animals must be sought in the heart, in blood-less animals in an analogous organ" (*Juv.*, Hett trans. 469b3 and 4).

Galen said that the part of the soul residing in the heart causes "boiling as it were of the innate heat." Galen — and the neo-Galenists of the Renaissance — thought the heart's heat concocts inspired air into vital pneuma. In the Middle Ages and in early Renaissance times, belief in the heat of the heart was conventional, in various forms almost universal. Harvey asserted that in the heart the blood "is restored to its erstwhile state of perfection. Therein, by the natural, powerful, fiery heat, a sort of store of life [*vitae thesauro*], it is reliquefied and becomes impregnated with spirits and (if I may so style it) sweetness [*balsamo*]" (*Motion of the Heart*, Franklin trans., chap. 8). In a later work, Harvey says the passage of blood through the lungs prevents the blood from "getting overheated and swelling up like honey and milk on the boil" (*Circulation of the Blood*, Franklin trans., 40). But Harvey believed that the blood heated the heart rather than vice versa, and was not sure why this heating occurred in the heart rather than elsewhere.

Earlier Descartes had mentioned three fires, one in the heart, one in the brain ("as from vinous spirit"), and one in the stomach ("as from green wood") (*Anatomical Excerpts*, AT 11:538 [see, on the date of this, Georges-Berthier, 1914, 48, note 2]). In the present passage, he retains the idea of a spontaneous warmth in the stomach but makes it less important than the heart's heat which he equates with life itself (see note 26). On the heat of the heart, see also the *Discourse*, AT 6:49–58; *Passions*, AT 11:330; *Description of the Body*, AT 11:280–282; and Rothschuh, D., 47, note 3. In letters between Descartes and Plemp (AT 1:497–499 and AT 1:521 ff. and 539), we learn that even in such cold-blooded animals as fishes there is greater warmth in the heart than elsewhere in the body. Descartes reexplains his theory of heart action in a letter to Beverwick (AT 4:4–6), and in part 2 of the *Description of the Body.*

The flesh of the lung is so rare and so soft, and always so re-
freshed by the air of respiration, that as blood vapors from the right
cavity of the heart come there through the artery that anatomists
call the "arterial vein" they are, in like measure, thickened and re-
converted into blood.[22] This blood then falls[23] drop by drop into
the left cavity of the heart where, if the vapors were to enter without
being thickened again, they would be inadequate to nourish the
fire[24] that is there.[25]

22. *Vaporization and condensation.* Commonplaces of everyday life, these can
occur on a small scale even without the application of flame (as when one warms
a glass of spirits in the hand). They occur more vividly in flame-induced distillation, familiar since ancient times (Aristotle mentions purification of sea water by
this method). The Arabs made extensive use of distillative extraction and improved
on ancient stills in various ways. Degree of rarefaction had also been a theoretical
problem for physics, going back at least to Anaximenes' doctrine that everything
consists of air compressed in different degrees.
  Descartes's formulation may owe something to Harvey, who defended calling the
blood flow "circular" by comparing it to the cycle of evaporation and condensation
that produces rain. Harvey said the blood may be coagulated in passing through the
body parts and reliquefied in the heart (see note 21). With Descartes, it is the lungs,
rather than the body parts in general, that recondense volatilized blood. In commenting on this passage, La Forge insists that Descartes meant not that the blood
is volatilized by the heart's heat but merely that it swells as milk does when it boils
(*l'Homme*, 187–188). For more on this subject, see the *Description of the Body*
(AT 11:236–237) and the references in Gilson, *ISC*, 50 and 51.

23. If the blood falls into the heart, it should prove disruptive to the circulation
for a man to be inverted. Descartes says that in such cases blood "flows" or is "inserted" into the heart "by the circulation and by the spontaneous contraction of the
vessels" (letter to Regius, November 1641, AT 3:440–441).

24. The idea that blood nourishes the fire in the heart is an adaptation of a similar analogy Galen drew in comparing the heart to a lamp; both require fuel (*UR*,
chap. 3, K4, 488 ff.). Georges-Berthier says "*La comparaison cartésienne du corps avec
une lampe (p. ex. [A-T] xi, 169) était banale à l'époque*" (1914, 49, note 3).

25. *Rarefaction and condensation in the activity of the heart.* According to Descartes, rarefactions in general are of two sorts, evaporative (*cum liquor in fumam
sive aërem abit & forman mutat*) and merely volumetric (*cum liquor formam retinet
& mole tantum augetur*). The latter sort may occur progressively (*sensim sensim*)
or instantaneously (*in momento*).
  The rarefaction occurring in the heart is of the instantaneous volumetric variety.
It is like fermentation in the sense that a small moiety can communicate its condition to a large one — as in making beer, wine, or bread. A little blood left behind
in the recesses of the heart after evacuation "acquires a new degree of heat and as it
were a fermentive character," which it communicates suddenly to new blood coming
in from the veins (letter to Plemp, AT 5:530–531). Note that Descartes likens —
but does not precisely equate — cardiac heating to fermentation. La Forge defends
Descartes's fermentative hypothesis of diastole at length against alternative hypotheses
such as that of a "pulsific faculty" of the soul — which no more explains the dilatation of the heart, he says, than an elephant is explained by calling it "un animal

And thus you see that breathing, which in this machine serves solely to thicken the vapors, is as necessary for supporting the fire in its heart as breathing is necessary in us for sustaining our life.[26] [Such is the case], at least, in those of us who are fully formed. For in infants still in their mothers' womb, and hence unable to draw in fresh air by breathing, two conduits make up the lack.

---

d'Afrique" (l'Homme, 183). For Harvey on this point, see note 21.

In comparing various sorts of vital actions with fermentation, Descartes — and in this he differs little from contemporaries interested in the subject — meant only to speak in a loose analogical way about what we should term thermogenic or exothermic reactions. Thus he often lumps fermentation with such reactions as that between quicklime and water, that between strong acid and iron, and the like. See letters of Descartes to Plemp and to Newcastle, April 1645, AT 4:189; also notes 14 and 26.

In characteristically corpuscular terms, Descartes revises the traditional view of fermentation. Late sixteenth-century alchemists (among them Bonus and Libavius) had assigned to fermentation an ennobling and organizing role. For these alchemists, the philosopher's stone itself had the character of a ferment. Moreover, they compared the "great work" to the replicative and organizing action of the Aristotelian nutritive soul.

Like these authors, Descartes makes fermentation central to life, but he does so in his usual corpuscularizing way. Emphasizing heat production rather than exaltation, he views the very beginning of life as a fermentive interaction occurring between male and female seminal fluids. Blood formation, too, is described as a kind of fermentive activity. But above all, fermentation generates the heat of the heart, and this heat activates the body in general.

I am grateful to Dr. W. Pagel for personally calling to my attention the alchemical theories of fermentation which were an undoubted source of Descartes on this subject, although he changed them almost beyond recognition. See Pagel, "Van Helmont's ideas on gastric digestion and the gastric acid," Bulletin of the History of Medicine, 30(1956), 533–536. Also, the many references to fermentation and fire in the indexes to the Theatrum chemicum, praecipuos selectorum auctorum tractatus de chemiae et lapidis philosophiae . . ., Strasbourg, H. Zetsner, 1659–1660; and in particular P. Boni (or G. Lacinio?), Praeciosa ac nobilissima artis chymiae collecteana . . ., Nuremberg, G. Heyn, 1554, trans. A. E. Waite, The New Pearl of Great Price . . ., London, V. Stuart, 1894, pp. 252–270, and A. Libavius, Alchemia recognita, emendata, et aucta, Frankfurt, Saurius, 1616, pp. 35–36.

26. Heat and life. The focal assumption of Descartes's physiology is that heat, and not the soul, is the primary cause of the physiological, as opposed to the cognitive, acts of man. He develops this thesis in the Discourse (AT 6:52–55), in the Passions (AT 11:329–331), and in a number of letters to Mersenne (for example, AT 3:122) and especially Henry More, to whom he writes: ". . . life I deny to no animal . . . except insofar as I lay it down that it consists simply of the warmth of the heart" (AT 5:267).

There were objections to this view and to the automatistic correlates that went with it (see those, for instance, which Fromondus expressed to Plemp, AT 1:403, and Descartes's rebuttal, AT 1:416, as well as the remarks on this subject in the Description of the Body, AT 11:227). The idea that the blood distributes an enlivening heat from the heart was not new in itself. Aristotle and Galen — and indeed most medieval and Renaissance medical thinkers — held generally similar views. Their ideas differed from Descartes's, however, in that most of them made the heat an instrument of the soul, whereas for Descartes the heat replaces the soul in all activities except cognitive ones.

Through one of these [the foramen ovale] blood passes from the vena cava to the vein-called-artery [or pulmonary vein], while through the other [the ductus arteriosus] the vapors or rarefied blood are breathed out from the artery-called-vein [the pulmonary artery] and go into the great artery [the aorta].[27] And as for animals that have no lung at all, they have but a single cavity in their heart; or if they have several, these are all in a single sequence.[28]

The pulse, or beating of the arteries, depends upon eleven small membranes[29] which like little doors close and open the orifices of the four vessels that open into the two cavities [ventricles] of the

27. *Fetal circulation.* Galen was aware of the changes that occur at birth and realized that they were adaptively associated with the commencement of breathing, the lung needing nourishment even before it begins to take in air. In Galen's scheme, both veins and arteries carried blood to the organs. He viewed the foramen ovale as transferring blood from the vena cava to the pulmonary vein for delivery to the lungs. The ductus arteriosus functions similarly, he supposed, transferring arterial blood from the aorta to the pulmonary artery — again, for delivery to the lungs (*UP*, bk. 15, chap. 6, K2, 443–445). It was Harvey who, building on Columbus' theory of pulmonary circulation, set these matters to rights (*Motion of the Heart*, chap. 6, 48–49 and 155). For Descartes elsewhere on fetal circulation, see his *Discourse*, AT 6:53, and *Description of the Body*, AT 11:237–238, where he states that in diving birds the shunts that bypass the lungs in the fetus persist after birth.

28. *Auricles.* Today we know the foramen ovale admits blood from right auricle to left. To Descartes the auricles were not so much parts of the heart as terminal distensions (*bourses*) of the main veins returning the blood to the heart (pulmonary vein and vena cava); see his *Discourse*, AT 6:49–50, and *Description of the Body*, AT 11:231 and 233. Galen had called these organs *ota* (ears) and described them as "hollow, fibrous epiphyses," or sometimes "apophyses," of the heart which push their contents into the ventricles at the time the ventricles are exerting their pull. As reservoirs, the auricles permit the ventricles to pull in materials violently without harming the tributary veins, which might otherwise be ruptured (*UP*, bk. 6, chap. 15, K3, 480–487).

Galen acknowledged that he was not the first to call these organs ears (*Anat. Adm.*, bk. 7, chap. 9, K3, 615–616). In pre-Cartesian Renaissance anatomy, the auricles were usually described as parts, or appendixes, of the heart (for example, by du Laurens, Crooke, Falloppius, Paré, Bauhin, Bartholin, Columbus, and Piccolhomini); but at least one author, Riolan, "would judge them rather to be little portions of the vessels to which they are attached" (*Oeuvres*, 548). And near the beginning of the fourth chapter of his *Motion of the Heart*, Harvey speaks of the contractions of the auricles as distinguished from those of the heart itself.

29. The eleven "membranes" are as follows: the left auriculoventricular (mitral or bicuspid) valve (two flaps), the right auriculoventricular (tricuspid) valve (three flaps), and the semilunars of the aorta and of the pulmonary artery (three flaps each). They are also enumerated in the *Discourse*, AT 6:47–48, and discussed in a letter of Descartes to Mersenne, AT 1:377–378, and again in the *Description of the Body*, AT 11:228–230. The number and distribution of the heart valves had been spelled out by Galen (*UP*, bk. 6, chap. 14, K3, 476–477), and often in early Renaissance physiological anatomy.

heart. For, at the moment when one beat ends and another is about to begin, the little doors at the orifices of the two arteries are tightly shut, while those at the orifices of the two veins are open; whence two drops of blood cannot fail to fall immediately from these two veins, one into each cavity of the heart.[30] Next, these drops of blood, being rarefied and suddenly occupying a space incomparably greater than before, push shut the little doors at the orifices of the veins, thus preventing more blood from descending to the heart, and push open those of the arteries — through which the vapors pass promptly and forcefully, thus making all the arteries of the body inflate at the same time as the heart.[31] But immediately thereafter, this rarefied blood either is condensed again [in the lungs] or penetrates other parts of the body; and thus the heart and the arteries are deflated, the little doors at the orifices of the two arteries are shut again, and those at the orifices of the two veins are reopened to admit two

30. Two major problems are here identified.

First, what causes the heart to empty? For Harvey, the cause resided in the contraction of the heart's muscular wall: "When the ventricle is full, the heart raises itself, forthwith tenses all its fibers, contracts its ventricles, and gives a beat. By this means it ejects at once into the arteries the blood discharged into it by the [two] auricle[s]" (Motion of the Heart, Franklin trans., 39).

For Descartes, the cause is fermentation which produces a vaporization of the blood; the vaporization dilates the heart and opens its valves to permit the escape of the blood. Later, he was to disagree with Harvey's idea that the heart chambers are reduced in volume at the time that the blood comes out (Description of the Body, AT 11:231–233). When a critic objected to fermentaton as a plausible cause of dilation and mentioned that extirpated hearts can continue to beat, Descartes replied, first, that once the heart starts to beat, little is required to maintain the rhythm; second, that some trace of blood remains in extirpated hearts; and third, that this residue interacts with a ferment-like humor still present in the recesses of the heart (letter to Plemp, AT 1:522–523). See also note 25 above.

Second, what causes the heart to fill? Descartes assigns a primary role to gravity (we keep hearing that "drops" of blood "fall" into the cavities of the heart). Harvey had said that the auricles "are replenished as a storeroom or reservoir [lacuna] would be, this occurring through a spontaneous diversion of the blood [declinante sponte sanguine] and through venous movement toward the center"; and that "the blood does not enter the ventricles through the attraction or distension of the heart but is sent in by a beat of the auricles" (retranslated from the Latin, Motion of the Heart, chap. 4; the passages appear in the Franklin translation on pp. 142–143).

31. Descartes uses the term "pulse" to refer to the arterial beat and sometimes also the heartbeat. He does not envision a pulse-wave, however (Description of the Body, AT 11:232). Since blood cannot enter the arteries without at the same time displacing blood already there, what he visualizes is an instantaneous expansion of the arterial blood as a whole. He reports experiments in which he incises an artery on the cardiac side of a ligature to show that the blood is driven, not sucked as Galen said, into the arteries from the heart (letter to Plemp, AT 1:526–527).

more drops of blood which cause the heart and arteries to be inflated again exactly as before. [Cl–6; AT–125]

Knowing the cause of the pulse, it is easy to understand that it is not so much the blood contained *in the veins* of this machine, newly come from its liver, as blood contained *in its arteries*, already distilled in its heart, that can become attached to other parts [that is, the parts of the body in general] and can serve to replace what the continual agitation of these parts — and the divers actions of neighboring ones — detaches from them and sends away.[32] For the blood *in the veins* flows ever little by little from their extremities toward the heart (and the arrangement of certain little doors or valves which the anatomists have noted in several places along the veins must also persuade you sufficiently that it happens in the same way in us); but per contra, blood that is *in the arteries* is pushed out of the heart, under pressure and in separate little thrusts, toward their extremities; whence this blood can easily come to join, and unite with, all the parts [of the body], and can thus maintain them or even make them grow if the machine represents the body of a man so disposed. [Cl–7; AT–126]

For at the moment when the arteries inflate, the particles of the blood they contain will here and there strike the roots of certain little threads, which, originating from the extremities of the little branches of these arteries, compose bones, flesh, membranes, nerves, the brain, and all the rest of the solid members according to the different ways in which they are joined or interlaced. These particles are strong enough to push the fibers before them somewhat and in this manner gradually to replace them. Then, at the moment when the arteries deflate, each of these part[icle]s stops where it is, and by that fact alone is joined to the particles it touches, in conformance with what has been said hitherto.[33]

32. In considering the flow of nutrients, Descartes means to reject the Galenic view that the liver suffices to ready blood for distribution to the tissues in favor of his own idea that the blood must first undergo a cycle of distillation (in the heart) and recondensation (in the lungs). The latter view squares better with Harvey's demonstration that blood cannot go directly from the liver to the extremities of the veins, as Galen had supposed, but must first pass twice through the heart, whence it proceeds to the body by way of the arteries only and never by way of the veins.

33. *Assimilation.* Descartes viewed the tissues as fibrous (see note 15). His interest in the composition of tissues, and in the process by which they incorporate additional substance, is representative of his desire to define form and function in

Now if it is the body of an infant that our machine represents, its matter will be so tender and its pores so easily enlarged, that the part[icle]s of the blood which enter thus into the composition of its solid members will generally be a little coarser than those whose places they take, or it will even happen that two or three together will replace a single one, and this will cause growth. But in the meantime, the matter of its members will harden little by little so that after a few years its pores will no longer be able to enlarge to the same extent; and so, ceasing to grow, the machine will represent the body of an older person.[34] [Cl-8; AT-127]

Moreover, only a very few blood part[icle]s would be able to unite each time with the solid members in the manner just described; rather, the majority return in the veins from the extremities of the arteries which in many places are joined to those of the veins. And some part[icle]s perhaps also pass out of the veins for the nourishment of some of the members; but the majority go back into the heart and from there go into the arteries again; whence the movement of the blood in the body is merely a perpetual circulation.

In addition, some part[icle]s of blood proceed to the spleen, and others to the gall bladder. And — from the spleen and the gall as

microstructural and microdynamic terms. His models are only guesses. If they seem gratuitous at times, we must remember that effective microscopy was still decades away. He was careful to obtain a certain license for free imagining by the device of choosing to describe not man but a hypothetical machine which man himself may or may not resemble. For more of Descartes's views on fiber replacement, see his *Description of the Body*, AT 11:246; also *Anatomical Excerpts*, AT 11:596–600, where he distinguishes "appositive" from "immutative" (intussusceptive) nutrition.

34. The association of age with dryness (and/or hardness) was traditional. Aristotle said that "the matter of which bodies are composed among the living consists of hot and cold, dry and moist. But as they grow old they must dry up" (*Long.*, Hett trans., 466b20 ff.). Galen said that "that which all men commonly call old age is the dry and cold constitution of the body" (*de San.*, bk. 5, chap. 9, trans. R. M. Green, *Galen's Hygiene*, 218).

During the Middle Ages, the idea developed that the body contains a certain "radical humidity" which, being — unlike the other parts — irreplaceable, gradually dries up. This idea persisted into the late sixteenth century; thus Paré: "Now in old age men are cold and dry, . . . [because of] the consumption of the radical or substantific humour proceeding from the multitude of years" (*Oeuvres*, 1585, bk. 1, chap. 9). For Fernel, a body engendered of blood and semen must begin by being hot and wet. Weighing whether maturation is primarily a cooling or a drying process, he decided that both are involved, although the drying process is conspicuous earlier (*Physiologie*, bk. 3, chap. 10).

For Descartes elsewhere on aging and hardening, see his *Description of the Body*, AT 11:250.

well as from the arteries directly — some part[icle]s reenter the stomach and the bowels where they act like aqua fortis to help the digestion of food.[35] And because these are brought here from the heart quasi-instantaneously, they are invariably very hot, which enables their vapors to rise easily through the gullet toward the mouth, there to compose the saliva.[36] There are also some that flow out as urine through the flesh of the kidneys and as sweat or other excrements through the skin. And in all these places it is only the position, or shape, or small size of the pores through which they pass that makes some go through rather than others and keeps the rest of the blood from following, as you may have seen in divers sieves which, being differently pierced, serve to separate different grains one from the other.[37] [Cl–9; AT–128]

But what must be chiefly noted at this point is that all the liveliest, strongest, and subtlest parts of this blood proceed to the cavities of the brain, inasmuch as the arteries that bring them there are the ones that come in the straightest line from the heart; and, as you know, all bodies in motion tend insofar as possible to continue moving in a straight line.

For example, observe the heart, A [in Fig. 1], and consider that, when the blood leaves it forcefully through opening B, all [the blood] part[icle]s tend toward C, that is, toward the cavities of the

Figure 1

35. Descartes proceeds, in his accustomed manner, to reformulate Galen's ideas in terms of Cartesian physics. Galen said that the gall bladder and spleen attract yellow and black bile, respectively, from the liver; that they alter and incorporate a part of those humors; and that they expel the unused portion. The residue from the gall bladder proceeds to the intestine; the residue from the spleen, to the stomach. Here the two humors variously aid the activities of those organs (see notes 12 and 121). In replacing the humors with particles, Descartes retains the mistaken idea of a flow from these two organs to the alimentary tract.

36. Galen and many neo-Galenic anatomists had better ideas than did Descartes about the origin of saliva. Galen said that saliva was generated by glands and poured into the mouth through visible ducts, but he did not assign a digestive role to this substance. In a slightly later letter to Mersenne, Descartes moves closer than in *Man* to Galen's position, acknowledging that saliva comes from the "almonds" (salivary glands), but he supposes that saliva tends to be swallowed unless muscularly forced into the mouth; he adds that saliva is sometimes supplied by the arteries of the gums (AT 3:139).

37. In this and the preceding paragraph, Descartes expresses his belief that the blood is composed of nutrients and substances destined for secretion and excretion. He supposes further that certain of its finer particles, when filtered off, become animal spirits (see note 43). In a letter to Mersenne (AT 3:139–140), he says that the particles of the gastric secretions, saliva, sweat, and tears slide out "like little eels" through the extremities of the arterioles, leaving behind the colored particles whose branching shape causes them to become interlocked and congealed.

brain; but the passage not being large enough to carry all of them thither, the weakest are turned back by the strongest, which thus proceed there alone.[38] [Cl–10; AT–128]

And it may be noted in passing that the strongest and liveliest particles other than those that enter the brain go to the vessels designed for generation. For example, if those that have the force to reach D cannot advance farther to C (because there is not room for all of them there), they turn instead toward E, rather than toward F or G, inasmuch as the passage toward E is straighter. Beyond which, I perhaps could show how, from the humor that assembles at E, another machine, quite similar to this, can be formed; but I do not wish to enter further into this matter.[39]

38. Here Descartes examines the dynamics of body fluids, utilizing the laws of motion that he will develop more systematically in part 2 of his *Principles*, AT 8:53–79 (9:76–102). Movement as such is God-given (article 36; see also *Light*, AT 11:11–12; and letter to Newcastle, AT 4:328). A body at rest or in rectilinear motion continues so unless something induces a change (articles 37 and 39). Impact, for example, may alter the speed or direction of one or both the bodies involved (articles 45–61). While movement is transferred in such cases, it is not created or destroyed; the total amount in the universe never changes (article 36; also *Light*, AT 11:13 and 36–48).

A partial set of rules is worked out, in the *Principles*, for collision. The results of collision depend (*a*) on the relative sizes (weights) and speeds of the bodies, (*b*) on whether one or both are in motion at the moment of impact, and (*c*) if both are in motion, on whether they are moving in the same or contrary directions (articles 45 ff.). Special details are furnished about the movements of particles in liquids (articles 56–58).

What Descartes says in the present passage about particle movement in the blood is consonant with, though not fully explained by, the rules he will develop later in his *Principles*. We cannot be sure how precisely he had formulated these rules when working on *l'Homme*. What is important, in any case, is the aim: to account for visible function in terms of a corpuscular dynamics common to living and nonliving things.

39. *Reproduction.* Descartes's machine "similar to this" is an offspring. He treats the subject of generation at length in his *Description of the Body* and in his *First Thoughts on the Generation of Animals*. The idea that the substrate of psychic activity must be related in some special way to the substrate of generation had recurred from time to time. Plato took over from Alcmaeon an idea that the soul-stuff passes from the brain via the spinal column to the urogenital tract whence it issues as seed-stuff (*Timaeus*, 90e–91d). A not dissimilar idea had appeared earlier in the Hippocratic treatise *On Generation* (*Oeuvres complètes*, Littré ed., Paris, 1851, vol. 7, pp. 470–543). Variants of this view had a number of Renaissance proponents; it was represented, for example, in the famous coition figures of Leonardo. Whether or not Descartes was influenced by the past history of ideas associating the psychic humor with the generative humor is unknown. For an account of the roots of the idea, see E. Lesky, 'Der enkephalomyogene Samenlehre,' "Die Zeugungs- und Vererbungslehren der Antike und ihr Nachwirken," *Abhandlungen der geistes-*

As for those parts of the blood that penetrate as far as the brain, they serve not only to nourish and sustain its substance, but also and principally to produce there a certain very subtle wind, or rather a very lively and very pure flame, which is called the "animal spirits"[40] For one must know that the arteries that bring blood from the heart, having divided into an infinity of little branches and having composed the little tissues that are stretched like tapestries at the bottom of the concavities of the brain [the choroid plexus], reassemble around a certain little gland [the pineal][41] situated near the middle of the brain's substance just at the entrance to its

---

und sozialwissenschaftlichen Klasse, Akademie der Wissenschaften und der Literatur in Mainz (Wiesbaden, 1950), 1233–1254.

The nub of Descartes's theory of reproduction is that "the seed . . . of animals, being very fluid and ordinarily produced by the conjunction of the two sexes, seems only to be a confused *mélange* of two liquors which, serving to ferment one another, are heated so that some of their particles, acquiring the same agitation as in fire, are spread apart and press against others and in this way dispose them gradually in the fashion requisite to form the parts" (*Description of the Body,* AT 11:252–253). About half of Descartes's *Description* is an elaborate account of corpuscular movements that give rise to the parts of the body in epigenetic sequence. The whole picture is speculative and deductive and exhibits the sometimes disastrous consequences of his incautious dependence on reason divorced from adequate observation.

40. *Origin of spirits.* Descartes's account is basically a "Cartesianized" adaptation of Galenic ideas that had been transmitted by medieval and early Renaissance anatomists without any fundamental change. Galen's opinion was that air is concocted in a preliminary way in the lungs and definitively in the heart; vital spirits, the product, travel to the brain in the blood and there give rise to animal spirits by a process of "exhalation" (*anathymiasis*) (*de Plac.,* bk. 8, chap. 8, K5, 709; *UP,* bk. 6, chap. 17, 496 and bk. 7, chap. 8, K3, 541; and *UR,* chap. 5, K4, 502 and 506). Descartes omits the vital spirits — the animal spirits preexisting, for him, as subtle particles of blood. He adapts the Galenic idea of *anathymiasis* to his own corpuscular suppositions by postulating a process of differential diffusion: spirit particles leave the blood through pores in the arterial walls of the choroid plexus or small vessels surrounding the pineal gland.

For Descartes's discussions of animal spirits, see Gilson, *ISC,* 99–103, who quotes some possible medieval sources of Descartes on this topic. See also Rothschuh, *D.,* 52, note 2; and notes 43, 44, and 128 below.

41. Thus far, we have not heard that the pineal gland is the seat of the soul; this news is reserved for a later part of our treatise. What is claimed for the gland here is that it is the destination of very subtle blood particles (animal spirits). The reasons for its receiving them will be developed extensively in later sections of *Man.* For Galen on the pineal gland, see note 42; and for Descartes on the pineal as the seat of the soul, see note 140; also Rothschuh, *D.,* 54–55, note 3.

In a posthumous anatomical note, Descartes says that the choroid plexuses are hung like tapestries around the pineal gland. Spirits rise (with blood) from the pituitary through the infundibulum and, if strong enough, reach the pineal gland; otherwise they are deflected toward the fourth ventricle and leave the brain via an opening near the optic chiasma (*Anatomical Excerpts,* AT 11:582–583).

cavities. [And one must know also] that the arteries in this region have many little holes through which the subtlest parts of the blood can flow into this gland, but which are so narrow that they refuse passage to larger particles. [Cl–11; AT–128]

It is also necessary to know that the arteries do not stop there but, being gathered several into one [the great vein of Galen], they rise straight up [through the straight sinus] and enter into that great vessel [the sagittal sinus and connections] by which like a Euripos the whole external surface of the brain is bathed.[42] It is further necessary to note that the coarsest parts of the blood can lose much of their agitation in the turnings of these little tissues through which they pass, inasmuch as they have the power to push the smaller ones that are among them and so to transfer to them some of their movement. The smaller ones, however, cannot lose their movement in this way, since [a] their agitation is augmented by that which the coarser ones transfer to them, and [b] there are no other bodies around them to which they themselves can as easily transfer theirs.

Whence it is easy to conceive that when the coarsest [particles in the blood] mount directly toward the external surface of the brain, where they serve to nourish its substance, they cause the smallest

42. *Choroid plexus.* Descartes like many before him made the pineal an *inward projection* into the brain cavity (from its floor) rather than an *outward* projection (from its roof). But his angiology is rather good. The choroid plexuses do drain, via the great veins of Galen, into a single vessel, the straight sinus, and from there into the general sinus system of the cortex. These structures had been described in detail by Galen, who believed, however, that veins (as well as arteries) bring blood *to* the choroid plexus. Galen thought that nature had provided the pineal gland as a support for veins carrying blood downward into and upward out of the choroid plexus (*UP,* bk. 9, chaps. 6–7, K3, 708–710). For Galen and pre-Cartesian anatomists on the pineal, see also note 135; for post-Cartesian developments, see Rothschuh, *D.,* 55, note 1.

In comparing the cerebral sinuses to the straits of Euripos, Descartes presumably means to speak in a general way without reference to the reversible tidal currents occurring there. He acknowledges the unidirectional flow of blood as specified by Harvey, who said that "blood does not follow Euripos and become unduly disturbed, moving hither and thither or even backwards, etc." (*Motion of the Heart,* chap. 7, Franklin trans., 55). Harvey attributes to André du Laurens the comparison of blood flow in the heart to the ebb and flow of Euripos (*Motion of the Heart,* chap. 1, 23). What du Laurens said was: "I think the nature of this motion no less worthy of admiration than the seven times daily reflux of the sea in the narrows of Euripos in the Euboean Strait. Aristotle, exiled in Chalcidos, being unable to give the cause of this phenomenon, was consumed with disappointment and died" (*Historia,* 1600, 472).

and most agitated [particles] to turn aside, and all of them to enter this gland, which must be imagined as a very full-flowing spring, whence they flow simultaneously in every direction into the cavities of the brain. And thus, without any preparation or alteration except that they are separated from the coarser ones and still retain the extreme rapidity that the heat of the heart has given them, they cease to have the form of blood and are designated animal spirits.[43] [Cl–12; AT–130]

Now in the same measure that spirits enter the cavities of the brain they also leave them and enter the pores [or conduits] in its substance, and from these conduits they proceed to the nerves. And depending on their entering (or their mere tendency to enter) some nerves rather than others, they are able to change the shapes of the muscles into which these nerves are inserted and in this way to move all the members. Similarly you may have observed in the grottoes and fountains in the gardens of our kings that the force that makes the water leap from its source is able of itself to move divers machines and even to make them play certain instruments or pronounce certain words according to the various arrangements of the tubes through which the water is conducted.[44] [Cl–13; AT–130]

43. *Animal spirits.* These are merely the smallest (least massive) and most agitated third-element (terrestrial) particles of the blood. Descartes's most precise and comprehensive statement on their nature is contained in a letter to Vörstius written more than a decade after the completion of *Man.* There we learn that spirits are intermediate in their constitution between two other third-element substances, namely fire and air (see note 4). The latter are alike in that their particles are moved along by direct contact with particles of the first and second elements (often spoken of by Descartes as "subtle matter"), fire moving faster under these conditions than air. He says that "all bodies consisting of terrestrial particles that [a] are bathed in subtle matter and [b] are more agitated than the particles of air, but less so than those of flame, can be spoken of as spirit[s]." He does not disallow the existence of spirit particles acquired with food; such particles are comparable to substances often termed "natural" and "vital" spirits. He even intimates, without clearly stating, that these particles may be turned into animal spirits. But he assigns them no physiological role of their own, and in his formal physiological works in this area, we hear nothing about them (AT 3:686–689).

44. Descartes here is considering animal spirits in relation to nervous action. The doctrine of a subtle fluid in nerves had been developed in antiquity in a preliminary way by Herophilus (late fourth century B.C.) and Erasistratus (early third century B.C.), and definitively by Galen. For early theories on nervous action, see F. Solmsen, "Greek philosophy and the discovery of nerves," *Musaeum Helveticum* 18 (1961). Galen saw spirits not as moving vectors of sensory or motor transmission, not as "impulses," but rather as a more or less persistent substrate for motive and sensitive soul-faculties — nerves being, for him, extensions of the brain. For a partial

And truly one can well compare the nerves of the machine that I am describing to the tubes of the mechanisms of these fountains, its muscles and tendons to divers other engines and springs which serve to move these mechanisms, its animal spirits to the water which drives them, of which the heart is the source and the brain's cavities the water main. Moreover, breathing and other such actions which are ordinary and natural to it, and which depend on the flow of the spirits, are like the movements of a clock or mill which the ordinary flow of water can render continuous. External objects which merely by their presence act on the organs of sense and by this means force them to move in several different ways, depending on how the parts of the brain are arranged, are like strangers who, entering some of the grottoes of these fountains, unwittingly cause the movements that then occur, since they cannot enter without stepping on certain tiles so arranged that, for example, if they approach a Diana bathing they will cause her to hide in the reeds; and if they pass farther to pursue her they will cause a Neptune to advance and menace them with his trident; or if they go in another direction they will make a marine monster come out and spew water into their faces, or other such things according to the whims of the engineers who made them. And finally when there shall be a rational soul in this machine, it will have its chief seat in the brain and will there reside like the turncock who must be in the main to which all the tubes of these machines repair when he wishes to excite, prevent, or in some manner alter their movements. [Cl–14; AT–132]

But to make you understand all this distinctly, I wish to speak to you first of the fabric of the nerves and the muscles, and to show you how from the sole fact that the spirits in the brain are ready to enter into certain of the nerves they have the ability to move certain mem-

---

discussion of the vexed question of Galen's views on nervous transmission, see Siegel, *Galen's System*, 192–195, and many references in Siegel's more recent *Galen on Sense Perception* (Basel and New York, 1970).

Most sixteenth- and early seventeenth-century anatomists confirmed Galen's general views on nervous action, confining their disputes to certain details. For example, du Laurens, departing further than his contemporaries from Galenic ideas, was uncertain as to whether nerves communicate both spirits and motive and sentient faculties, or only the latter. In the above paragraph and those that follow, Descartes retains the idea that animal spirits are present and brings their relation to nervous action into line with his own mechanical views. See also note 84.

bers at that instant.[45] Then, having touched briefly on breathing and other such simple and ordinary movements, I shall tell how external objects act upon the sense organs. After that I shall explain in detail all that happens in the cavities and pores of the brain, what pathway the animal spirits follow there, and which of our functions this machine can imitate by means of them. For, were I to begin with the brain and merely follow in order the course of the spirits, as I did for the blood, I believe my discourse would be much less clear.

Observe [in Fig. 2], for example, nerve A whose external membrane is like a large tube containing several other small tubes, b, c, k, l, and so on, composed of a thinner, internal membrane; and observe that these two membranes [outer and inner] are continuous with the two, K [pia] and L [dura], that envelop the brain MNO.[46] [Cl–15; AT–133]

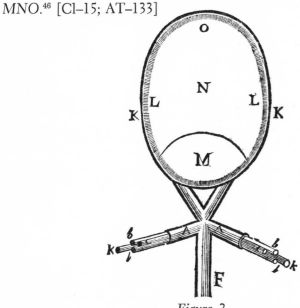

Figure 2

45. Does Descartes envision a traveling, time-consuming motor impulse? Probably not. The passage suggests that the mere transfer of spirits from the brain ventricles to the orifices of nerves within the brain suffices to cause a simultaneous outflow of spirits from the peripheral ends of those nerves. How the peripheral discharge of spirits permits the action of muscles is explained later on (see the text commented upon in notes 54, 58, 60, and 62).

46. The idea of the continuity of nerve coats with brain coats, like so many concepts that Descartes develops, comes from Galen who said that "each of the nerves growing out from there [the head] has a triple nature: its center and inmost part, analogous to the heart-wood of a tree, takes its beginning from the brain, and it is

Observe also that in each of the little tubes there is a sort of marrow composed of several very fine fibrils which come from the actual substance of the brain N and whose [two] extremities end [one] at the internal surface of the cavities of the brain and [the other] at the membranes and flesh on which the tubule containing them terminates. But because this marrow is not used to move the members, it will suffice for now that you know that it does not completely fill the tubes containing it but leaves room enough for animal spirits to flow easily through them from the brain into the muscles whither these little tubes, which should be thought of as so many little nerves, make their way.[47]

Next observe [in Fig. 3] how the tube or little nerve *bf* proceeds to muscle D, which I assume to be one of those that move the eye, and how it there divides into several branches composed of a loose membrane which can extend, enlarge, and shrink according to the quantity of animal spirits that enter or leave it, and whose branches

---

surrounded in a circle by an offshoot first of the pia mater, and second of the dura mater" (*de Plac.*, De Lacy trans., bk. 7, chap. 3, K5, 602; see also *de Loc. aff.*, bk. 1, chap. 6, K8, 57). Referring to spinal nerves, Galen further stated that "the 'pith,' so to speak, of each nerve branches off from the spinal medulla, and the meninges [of the brain and medulla] lie in a circle around it" (*de Plac.*, bk. 7, chap. 8, K5, 646). Descartes seems to base his reconstruction of the Galenic picture of nerve structure as much on Galen himself as on medieval or Renaissance modifications thereof.

47. *Sensory and motor functions.* We are about to hear Descartes place them both in one and the same nerve. In this, he continues sixteenth- and early seventeenth-century opposition to the Galenic theory that envisioned two sorts of nerves, sensory or soft (deriving from the cerebrum) and motor or hard (deriving from the cerebellum) (*de Plac.*, bk. 7, chap. 5, K5, 621–622 and *UP*, bk. 8, chap. 6, K3, 636–639). The distinction thus drawn in Galen's system was not an absolute one, but allowed for intermediate nerves as well, in which the harder the nerve the more predominantly motor its function. Galen distinguished further between the nerves of the several senses and the nerves of pain (*UP*, bk. 5, chap. 9, K3, 378). Vesalius followed Galen in acknowledging three "goals of nature" in the distribution of nerves, namely *motion, sensation* (in nerves attached to appropriate organs), and *awareness* of discomfort or distress (*ad tristantium dignotionem, Fabrica*, 417; see also note 116 below).

The Galenic-Vesalian theory had many advocates but was also challenged by a succession of theorists including Piccolhomini (*Praelectiones*, 261–262), du Laurens (*Historia*, 1600, 161, and *Oeuvres*, 1621, 102–103), Bauhin (*Theatrum*, 1605), and Bartholin (*Institutiones*, 1611, 395) who, arguing partly from the evidence of mixed nerves, insisted that the role of nerves in general is determined not by hardness or softness but by the peripheral organ into which they are inserted. Thus the way was paved for Descartes's departure from the view of separate nerves for sensory and motor function. His functional interpretation of the tripartite structure of nerves is given also in the *Dioptrics* (AT 6:109–112).

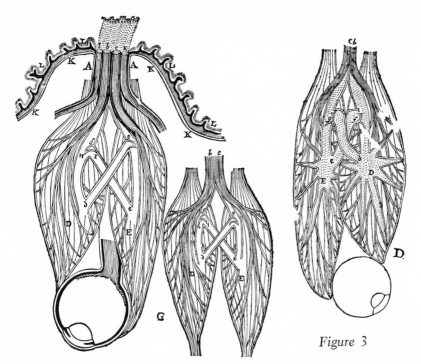

Figure 3

or fibers are so arranged that when animal spirits enter therein they cause the whole body of the muscle to inflate and shorten and so pull the eye to which it is attached; while on the contrary, when they withdraw, the muscle disinflates and elongates again. [Cl–17; AT–134]

Observe further that in addition to the [incoming nerve-]tube *bf* there is still another, namely *ef*, through which the animal spirits can enter muscle *D*, and another, namely *dg*, through which they can leave it. And quite similarly that muscle *E*, which I asssume is used to move the eye in the contrary direction, receives animal spirits from the brain through [nerve-]tube *cg* and from muscle *D* through *dg*, and sends them back toward *D* through *ef*.[48] And con-

48. *Reciprocal muscle action.* Although Galen had studied this subject and although his ideas won acceptance in Renaissance times, Descartes's view of the mechanism involved is distinctly his own. Contraction (swelling and shortening) is caused, he believes, by a double influx of spirits into the hollow, soft-walled nerve endings that partly constitute the muscle; some spirits enter from the muscle's own nerve, although not in copious amounts; others flow across a shunt from the antagonist, causing the latter to relax (deflate and elongate). The spirits from the muscle's

sider that although there is no evident passage through which the spirits contained in muscles D and E can leave them except to go from one to the other, nevertheless because their particles are very small and indeed because they are made incessantly finer through the force of their agitation, some always escape[49] across the membranes and flesh of the muscles while others return through the two [nerve-]tubes bf and cg [to replace those that escape]. [Cl–18; AT–135]

Finally, observe [Figs. 3 and 4] that at the juncture of the two tubes bf [muscle D's own nerve] and ef [the shunt from the other muscle], there is a certain small membrane Hfi that separates these two tubes and serves as a door. It has two flaps, H and i, arranged as follows. First suppose that animal spirits tending downward from nerve b toward flap H are stronger than those tending upward from muscle E toward flap i. [In this case] the descending spirits push down upon and open the membrane, permitting a prompt flow toward D of spirits [across the shunt] from muscle E. But [suppose on the contrary] that spirits tending upward [across the shunt from muscle E] are stronger than, or even only as strong as, those [tending downward from the nerve]. They would raise and close Hfi and thus prevent themselves from leaving E.[50] Whereas, if spirits are not strong enough to push the membrane from either side, it is its nature to stay open. Finally, [observe] that if, as sometimes happens, there is a tendency of spirits from D to go back through dfb

---

own nerve make their way in by pushing a valve (fH). The valve has another flap, fi, which is pulled open and permits a simultaneous inflow, across the shunt, from the relaxing antagonist.

For a discursive defense of Descartes's ideas on muscle action — especially the idea that spirits flow from relaxant to contractant — see La Forge's commentary (*l'Homme*, 224–259). For Galen on muscle antagonism, see *UP*, bk. 3, chap. 16, K3, 261, and esp. *de Mot. mus.*, bk. 1, chap. 4, K4, 384–396. Descartes seems not to have been influenced significantly by interim (medieval or Renaissance) elaborations of Galen's ideas on this subject. He was probably affected by the general acceptance of muscle reciprocity, but on this subject as on most he failed to specify his sources.

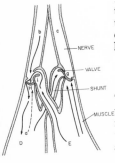

49. Descartes means that the system has a certain leakage.

50. Here Descartes stipulates the conditions under which reciprocal action occurs. The expansive force of spirits in E ordinarily pushes upward against and closes the valve Hfi, and the spirits are thus prevented from leaving. But if spirits entering D from its nerve press downward on this valve with force enough to keep it open, then the spirits in E can move across the shunt into D. The operation of the valves is such as to prevent a backflow of these spirits or their escape into the nerves.

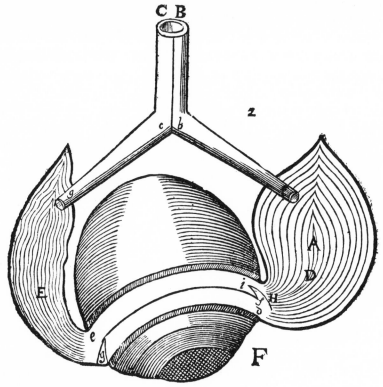

Figure 4

[into its nerve] or [back across the shunt] *dfe*, the flap *H* can stretch and thus block their passage. And similarly [observe] that between the two tubes *cg* [the other muscle *D*'s own nerve] and *eg* [the shunt from *E* to *D*] there is a membrane or valve, *g*, corresponding to *Hfi*. This membrane like its counterpart *Hfi* stands naturally ajar, but it can be opened by spirits coming in from [that muscle *E*'s own nerve-]tube *cg* and can be closed by those that tend to come in [from muscle *E*] by way of *dg*. [Cl–19; AT–136]

Whence it is easy to understand that if the animal spirits which are in the brain tend to flow only a little or not at all through the [nerve-]tubes *bf* and *cg*, the two little valves *Hfi* and *g* remain ajar and thus the two muscles *D* and *E* are lax and inactive; [this is true] inasmuch as the animal spirits that they contain pass freely from one muscle into the other, coursing from *e* through *f* toward *D* and reciprocally from *d* through *g* toward *E*. But if the spirits in the brain tend to enter forcefully into the two tubes *bf* and *cg*, and this force is equal on both sides, they forthwith both close the two passages *g* and *f* and inflate the muscles *D* and *E* as much as possible, in this way making them arrest the eye and hold it firm in the position it is already in.[51] [Cl–21; AT–136]

But if these spirits that come from the brain tend to flow with more force through *bf* than through *cg*, they close the little membrane *g* and open *f*, and this to a greater or lesser degree as they strike it more or less strongly. By this means, the spirits contained in muscle *E* proceed to muscle *D* through channel *ef*. They do so with greater or less rapidity as the membrane *f* is more or less open, with the result that muscle *D*, which spirits cannot leave, shortens while *E* elongates; and thus the eye is turned toward *D*. While, on the contrary, if the spirits that are in the brain tend to flow with more force through *cg* than through *bf*, they close the little membrane *f* and open *g*; whence the spirits of muscle *D* repair immediately by channel *dg* into muscle *E*, which by this means is shortened and draws the eye to its side.

For you well know that these spirits, being like a wind or a very subtle flame, cannot but flow promptly from one muscle into the other as soon as they find some passage, even though no other power propels them than that inclination which they possess to continue their movement according to the laws of nature. And you know besides that although they are very mobile and subtle, they lack not the strength to inflate and tighten the muscles in which they are enclosed, even as the air in a ball hardens it and stretches the skins that contain it.

---

51. Descartes recognizes that motionlessness of the muscle is of two sorts: *relaxed passivity* when neither antagonist is stimulated, and *tension without motion* when both are stimulated equally. In the next paragraph he attributes motion to differential activity of the two antagonists. The system is a modification of one put forward by Galen and here adapted, not very convincingly, to Descartes's corpuscular theory. Galen thought that the natural state of muscle was the contracted state (because the two ends of a severed muscle contract and because if a tensor is severed, flexion results from the unopposed contraction of the flexor). If muscles are motionless, Galen said, it is because the contractile tendencies of the antagonists are balanced. In volition, psychic faculties can alter this balance by reinforcing the contractile predisposition of just one antagonist so that it contracts at the expense of the other (*de Mot. mus.*, bk. 1, chap. 4, K4, 382–387, and esp. bk. 1, chap. 8, K4, 401–407). Accounts of muscle action by Descartes's immediate forerunners and contemporaries appear not to have added to Galen's ideas any new developments that significantly influenced Descartes.

Now it is easy for you to apply what I have just said of nerve A and the two muscles D and E to all other muscles and nerves; and so to understand how the machine about which I am telling you can be moved in all the ways that our body can, solely by the force of the animal spirits which flow from the brain into the nerves; because for each movement and for its opposite, you can imagine two little nerves or tubes such as bf and cg and two others such as dg and ef, and two little gates or valves such as Hfi and g. [Cl–22; AT–137]

And as for the ways in which the tubes are inserted into the muscles, although they vary in a thousand fashions, it is nevertheless not difficult to judge what they are by ascertaining what anatomy can teach you of the external form and use of each muscle.

Assuming, for example, that the eyelids [Fig. 5] are moved by two muscles of which one, namely T, serves only to open the upper lid and the other, V, serves alternately to open and close both lids, it is easy to suppose that these muscles receive spirits through two tubes, such as pR and qS, and that one of these, pR, proceeds to both muscles and the other, qS, to only one of them. And finally [it is easy to suppose] that the branches R and S, though seemingly inserted in the same fashion into muscle V, nevertheless produce quite opposite effects on account of the different arrangements of their branches or fibers.[52] This will suffice to make you understand

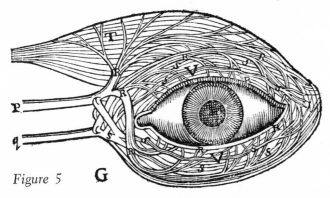

Figure 5   **G**

52. The interaction of the eyelid muscles is an elaborate affair. Descartes has grasped the fundamental feature, however, namely that the *levator palpebrae superioris* and the *obicularis oculi* are antagonists. In one of several treatments of the subject, Galen postulated two muscles — one in the wide (medial) angle of the eye tending to lower the upper lid, another in the narrow (lateral) angle tending to raise it. (*UP*, bk. 10, chap. 9, K3, 804–807). It is difficult to be sure what Galen

other [neuromuscular arrangements] as well. [Cl–23; AT–138]

Nor indeed is it difficult to suppose from the foregoing that the animal spirits can cause movements in all members in which nerves terminate, although there are some in which anatomists have found no nerves visible: such as the pupil of the eye, the heart, the liver, the gall bladder, the spleen, and other like organs.[53]

Now in order to understand in detail how this machine breathes, imagine that muscle d is the one that serves to raise its chest or to lower its diaphragm, and that the muscle E is its opposite; and that the animal spirits that are in the brain cavity marked m flowing through the pore or little channel marked n, which naturally remains always open, proceed first to tube BF where, lowering the little membrane F, they cause spirits to come from muscle E and inflate muscle d[54] [in Fig. 6].

---

really discerned (see May, *Galen*, 2:486–487). Vesalius, likewise, identified two muscles (the ones that exist in fact); he recognized that one of them was an encircling muscle and that this one tended to lower the upper eyelid, whereas the other tended to raise it (*Fabrica*, 238). That raising and lowering result from the reciprocal activities of antagonists had thus been specified by both Galen and Vesalius.

53. Actually, Vesalius mentioned nerve connections to the heart (*Fabrica*, 588), liver (506–507), gallbladder (509), and spleen (512). Descartes also, elsewhere, acknowledges nerves in all these organs (except possibly the liver). Heart nerves were generally recognized throughout the century preceding *l'Homme* (see for example Falloppius, *Opera*, 446; Piccolhomini, *Praelectiones*, 272; du Laurens, *Historia*, 467; Bauhin, *Theatrum*, 413–414), although there was some disagreement with respect to their role. Disregarding Fernel's Aristotelian idea that sentient faculties are extended by these nerves *from* the heart *to* the brain and other regions, the main point at issue was whether the heart beats spontaneously (in which case the nerves' function is primarily sensory) or at the command of the brain (in which case the heart's nerves extend to it a pulsific faculty originating in the brain).

54. *Muscular aspects of breathing.* On this, Galen said: "The thorax governs its [the lung's] dilatation and contraction and is [itself] moved by muscles and articulated by bones" (*de Caus. resp.*, K4, 465). Descartes correctly understands that inhalation involves a muscular raising of the ribs and tightening of the diaphragm. The responsible muscles can contract only if spirits flow in from an antagonistic muscle, so he invents a valve f which will permit this flow to take place. The machine is thus enabled to inhale.

But something must automatically bring an end to this process, and Descartes supposes that this must be the pressure that builds up during contraction. At a certain point, that pressure prevents the further entrance of spirits from the nerve, and the spirits must find a new path. They enter the antagonists and open the valve that reverses the intermuscular exchange; as a result, the automaton exhales.

This largely imaginary model invites comparison with the experimentally tested models of today. We still see exhalation as a matter of stopping the flow of stimuli to the muscles of inhalation. For us what flows is not spirits but a propagated disturbance (although the direct stimulus to the muscle has become a substance once again). A buildup of pressure is, for us as for Descartes, the cause of the necessary

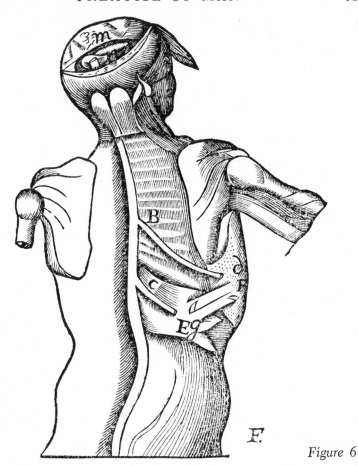

Figure 6

Next, reflect that around muscle *d* are certain membranes that press in on it more and more as it is increasingly inflated. The arrangement is such that, before all the spirits can pass from *E* to *d*, [the resistance caused by] these [membranes] stop[s] the flow [of spirits], causing them to be regorged, as it were, through *BF*. The result is that those from [nerve-]channel *n* are redirected. In this way, spirits proceed to *cg* and simultaneously force it open, causing muscle *E* to inflate and muscle *d* to deflate. This they continue to do as long as they are subject to the impetuosity of spirits tending

---

inhibition. However, we view that pressure as signalling the brain through afferent branches of the vagus nerve. A major difference between our model and Descartes's is that we view breathing out as primarily passive, as due to relaxation. The most striking likeness between the models is that both are based on reciprocal regulative feedback.

to leave muscle *d* under pressure from the membranes that surround it; and when this impetuosity is spent, they resume their course through *BF*. Thus, ceaselessly the two muscles are made to inflate and deflate in turn. You must suppose this to be true as well of other muscles that serve the same end [breathing] and must reflect that they are so arranged that when muscles like *d* are inflated the space which contains the lungs is enlarged, and this causes air to come in as into a bellows when one expands it; and that when muscles contrary to *d* [are inflated] that space shrinks, and this causes the air to leave again.[55] [Cl–25; AT–140]

To understand, as well, how this machine swallows the food that is at the back of the mouth,[56] assume that

[*a*] muscle *d* [in Fig. 6] is one of those that raise the root of its tongue and *hold open* the passage through which the respired air must pass in order to enter the lung; and that

55. *Aerodynamics of breathing.* Descartes's idea is one of a series of classic accounts. In antiquity, varying emphasis was placed on (*a*) the pores of the skin, (*b*) the nostrils, and (*c*) the lung cavities as portals of intake and output. Empedocles thought air enters not only through the nostrils but through the pores into which it is sucked when blood retires from the body-surface to the heart, and through which it is forced out when the blood returns from the heart to the surface (H. Diels, *Fragmente der Vorsokratiker*, B 100). Plato envisioned a reversible cycling of air and fire first out through the nostrils and in through the pores, then out through the pores and in through the nostrils (*Timaeus*, 79a). Aristotle discussed breathing in several contexts, used the bellows analogy (but pointed out its limitations), and emphasized that air must get all the way to the chest, not just in through the nostrils (*Resp.*, 473a15–474a24); he saw breathing as a mechanical function involving work against an external resistance (*MA*, 700a20–25).

Galen had about as good a picture of respiratory aerodynamics as Descartes, possibly better. He said, "The thorax governs its [the lung's] dilatation and contraction and is [itself] moved by muscles and articulated by bones" (*de Caus. resp.*, K4, 465). Insufflation is due to an enlargement, exsufflation to a contraction of the lungs brought about by a muscularly induced corresponding increase or decrease in the size of the thorax (*UP*, bk. 7, chap. 5, K3, 525 ff.; *de Mot. mus.*, bk. 2, chap. 9, K4, 458 ff.). Galen thought air also entered invisible stomata (pores) at the termini of arteries and that once inside, it could be converted into vital pneuma.

Comparisons between lungs and bellows had been common ever since the Greeks gave the latter a name (*physeron*) cognate with their word for breathing or blowing (*physarion*). The lung-bellows analogy appears in Fernel (*Physiologie*, 1542, 659–660). Thus, here as so often elsewhere, Descartes arrives at his interpretation by giving a nonpsychistic, corpuscular reinterpretation of ideas selected from earlier theorists.

56. Descartes explains, in this paragraph on swallowing, the mechanism by which the passage to the lung is closed while food is passing to the stomach, and the passage to the stomach is closed while air is passing to the lung. Unfortunately, there is no figure to accompany his explanation; Clerselier suggests to the reader that he "utilize the foregoing figure [6] and let the imagination supply what is wanting."

[*b*] muscle *E* is its opposite and serves to *close* this passage and by so doing open that through which the food in the mouth must descend to enter the stomach — or [serves] rather to raise the point of the tongue which pushes [the food] thither; and that

[*c*] the animal spirits that come from the cavity *m* of its brain through the pore or little canal *n*, which normally remains always open, proceed directly into tube *BF*, by which means they cause muscle *d* to inflate; and that

[*d*] this muscle remains thus always inflated as long as there is at the back of the mouth no food to press it; but that

[*e*] this muscle is so arranged that, when [food] is there, the spirits the muscle contains are immediately regorged through tube *BF* and cause those that come through channel *n* to enter through the tube *cg* into muscle *E* to which spirits coming from muscle *d* proceed as well; and thus finally that

[*f*] the throat opens and the food descends into the stomach, after which the spirits from canal *n* immediately resume their flow through *BF* as before. [Cl–26; AT–141]

From this example, you can also understand how this machine can sneeze, yawn, cough, and make the movements necessary to reject the various excrements.

To understand, next, how external objects that strike the sense organs can incite [the machine] to move its members in a thousand different ways: think that

[*a*] the filaments (I have already often told you that these come from the innermost part of the brain and compose the marrow of the nerves)[57] are so arranged in every organ of sense that they can very easily be moved by the objects of that sense and that

[*b*] when they are moved, with however little force, they simultaneously pull the parts of the brain[58] from which they come, and

---

Galen had developed an elaborate theory on this subject, a central feature of which was that the epiglottis is movable and closes off the pharynx during the act of swallowing (see, for example, UP, bk. 7, chaps. 16–19, K3, 585–594). Galen's theory is too complex to permit consideration here; it is based on an exhaustive and thoughtful anatomization of the muscles and cartilages of the mouth, pharynx, and larynx. As it stands, without a figure, Descartes's scheme is by contrast too deductive, too naively fitted to his mechanical prejudices, and too neglectful of details.

57. For Descartes's ideas on nerve structure, see notes 46 and 47.

58. *Sensory transmission.* Descartes does not depict such transmission in terms of a traveling or time-consuming impulse. In the paragraph that follows we shall hear

by this means open the entrances to certain pores in the internal surface of this brain; [and that]

[c] the animal spirits in its cavities begin immediately to make their way through these pores into the nerves, and so into muscles that give rise to movements in this machine quite similar to [the movements] to which we [men] are naturally incited when our senses are similarly impinged upon.[59] [Cl–27; AT–141]

Thus [in Fig. 7], if fire A is near foot B, the particles of this fire (which move very quickly, as you know) have force enough to displace the area of skin that they touch; and thus pulling the little thread cc, which you see to be attached there, they simultaneously open the entrance to the pore [or conduit] de where this thread terminates [in the brain]: just as, pulling on one end of a cord, one simultaneously rings a bell which hangs at the opposite end.[60] [Cl–28; AT–142]

---

him suggest instantaneous mechanical transmission. His theory is, in essence, a mechanical analogue of the prevailing idea of the day, namely that the nerves extend to the sense organs certain sensory faculties whose center is in the brain.

This theory as taught in the late sixteenth and early seventeenth centuries was not much advanced over that introduced by Galen, who seems to have assumed an act of conscious awareness occurring in the brain simultaneously with a sensory act at the surface — skin, eye, ear, and so forth (see, for example, de Plac., bk. 7, chap. 7, K5, 641–642; also Siegel, note 44 above). Shortly before the time of Descartes, questions began to be raised, especially in connection with visual sensation, that pointed the way toward a traveling wave or disturbance. Kepler, for example, inquired whether the "visive spirits" that establish continuity between eye and brain convey the image somewhat as ripples spread through water (Dioptrics, Prop. LXI). Descartes abides by the notion that the peripheral and central acts occur simultaneously.

59. For Descartes on motor transmission, see note 45.

60. Descartes does not use the term "stimulus" nor, in our sense, the term "response." He usually employs the word "action" to denote the elicitive events as well as the consequent events in the nerves. In this and the two succeeding paragraphs, he builds a mechanical model to explain them. His account may owe something to existing ideas on the subject.

The thinker whose scheme Descartes's most resembles is Bartholin, who stipulated the following sequence: (a) the object is perceived by an external sense, (b) appetition ensues, depending on intellectual adjudication of the sense perception, (c) appetition "strikes" (ferit) the brain, (d) the brain arouses (ciet) a nerve, (e) the nerve arouses a muscle, and (f) the muscle lengthens or shortens (Enchiridion, 835–836). As Descartes develops the story, consciousness is neither included nor explicitly excluded. He acknowledges that for his model to work, the activation of a particular bundle of nerve fibrils must lead to an outflow of spirits by way of particular nerves (see notes 62, 130, and 147–149). He defers discussion of that part of the problem, however, and deals first with its sensory aspects, especially sensory discrimination.

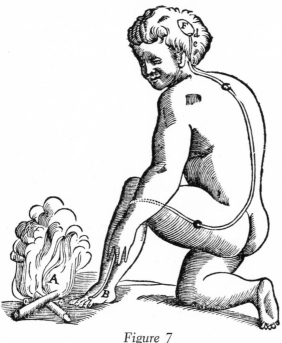

Figure 7

Now the entrance of the pore or small conduit *de*, being thus opened, the animal spirits from cavity *F* enter and are carried through it — part into the muscles that serve to withdraw this foot from the fire, part into those that serve to turn the eyes and head to look at it, and part into those that serve to advance the hands and bend the whole body to protect it.

But they can also be carried through the same conduit *de* into many other muscles.[61] And before stopping to explain more precisely in what way the animal spirits follow their course through the pores of the brain and how these pores are arranged, I wish to speak to you now in particular of all the senses that exist in this machine and to tell you how they are related to our own.

Know first, then, that a great many filaments like *cc* begin to separate one from another as soon as they arise at the internal sur-

61. The motions caused in the filaments of a particular nerve affect the brain in such a way that the animal spirits leave both by the same nerve and by others. Further details follow. See also Rothschuh, *D.*, 69–70, note 4.

face of the brain [of the machine] and that proceeding to spread thence through the rest of its body they serve there as an organ of touch. For although external objects do not ordinarily touch these filaments, but touch the skin that surrounds them, there is no more reason to think of the skin as the sense organ than to think of gloves as sense organs when we feel something while wearing gloves.[62] [Cl–29; AT–143]

And note that although the threads I speak of are very thin, yet they extend safely all the way from the brain to parts that are farthest therefrom, nor is there anything in between that breaks them or that prevents their activity through pressure, even though the parts are bent in myriad ways: because [a] the threads are enclosed in the same tubules that carry the animal spirits to the muscles, and [b] these spirits, always somewhat inflating the tubes, protect the fibers against crowding and keep them always maximally taut all the way from the brain whence they arise to the places where they terminate.[63]

And now I assert that when God will later join a rational soul to this machine,[64] as I intend to explain further on, He will place its

62. Nerve filaments arising close together at the surface of the brain ventricles accompany each other a short distance within the brain substance and then separate and reassemble into new bundles, each of which is projected beyond the brain as the central, filamentous "marrow" of a particular cranial nerve. The nerve walls, we remember, are extensions of the meninges of the brain. For Descartes's theory of sensory perception, see the *Principles*, AT 8:315–323 (9:310–317) and the *Dioptrics*, AT 6:109–114. See also notes 72 and 133 below.

63. One of the things Descartes wants his model to explain is the structural integrity and functional independence of each conductile nerve pathway despite its great length in comparison to its diameter (see also *Dioptrics*, AT 6:111–112).

64. *Body and soul.* Here we have for the first time in *Man* a serious consideration of the soul and the soul's relation to the body. Soul-body relations had been debated in the patristic and Scholastic literatures familiar to Descartes as a result of his Jesuit education. It is thus primarily in religious and only secondarily in medical documents that he finds the soul-body problem posed by others, even though he develops it here in a physiological context. In religious thought the classic purpose had been to reconcile Greek theories of the soul with ecclesiastical doctrine on the subject. It is not incorrect to think of the soul-body problem as having been eloquently posed in antiquity by Plato and as reaching a kind of culmination in the solution proposed by Descartes.

In the *Timaeus* (32c–37c), Plato described the separate construction by a divine Demiurge or Craftsman first of the body of the universe and then of its soul. Plato intended this story less as an account of what actually happened than as a way of helping his readers to understand the way things actually are — much as a teacher helps his students to understand a geometric figure by the pedagogic device of "con-

chief seat in the brain and will make its nature such that, according to the different ways in which the entrances of the pores in the internal surface of this brain are opened through the intervention of the nerves, the soul will have different feelings.[65]

Thus, firstly, if the filaments that compose the marrow of these nerves are pulled with force enough to be broken and thus are separated from the part to which they were joined, so that the structure of the whole machine is somehow less intact, the movement they then cause in the brain will cause the soul (to which it is

---

struction" (see T. S. Hall, "The biology of the Timaeus in historical perspective," *Arion*, 4 [1965], 109–122). Descartes, like Plato, "constructs" the body separately from the soul for reasons that are at least partly pedagogic: namely, to show the reader how much the body can do entirely on its own.

For both authors man comprised a body and, in addition, an ontologically separate soul. But the two men had different understandings of the duality thus proposed. Descartes, in *Man* and elsewhere, insists that the soul's only functions are cognitive ones: conscious perception and volition, memory, and especially reason. For Plato, the soul's functions included in addition to *nous* or mind (which he saw as residing in the head), passion (*thymos*) in the breast, and an appetitive-vegetative function (*epithymetikon*) between the midriff and the navel. For Plato as for all the Greeks, the soul was a general causal agent responsible for life as a whole.

As to the union between the soul and the body, Plato asserted that the soul was bonded or "moored" to the polygonal elementary particles composing the bodies both of the cosmos and of individual microcosms or men. For Descartes, the locus of connection is the pineal gland where the soul and the animal spirits affect each other mutually. We shall hear about their interaction in the passages to follow. He is critical, in this connection, of an early metaphor, re-expressed in medieval times, comparing the soul's relation with the body to a pilot's relation with his ship. He objects to this as too loosely analogical a manner of looking at the problem and as suggesting too indefinitely the intimate interactions that occur between the soul and the animal spirits (see his *Discourse*, AT 6:59; the *Sixth Meditation*, AT 7:71; and many other references enumerated by Gilson, *ISC*, 297–306). The pilot analogy to which Descartes objects perhaps originated with Aristotle, who said that the soul uses heat in assimilating nutriment to the body much as the helmsman uses his hand in steering the ship (*de An.*, 416b16–20). See also notes 2 and 158.

65. Descartes here introduces the subject to be treated in the paragraphs that follow, namely the physical basis of the soul's different feelings or sensations. The basic principle of his theory of sensory discrimination is that different kinds of particle motion (we should term these "stimuli") displace in different ways the outer tips of the filaments of the nerves, and that the resulting peculiarities of peripheral motion of the fibers give rise to corresponding peculiarities of motion in the brain. What happens in the brain, we shall hear, is that the pores or conduits through which the spirits leave the ventricles to enter the nerves are altered (enlarged, diminished, reoriented) in various ways. These changes in the pores give rise to changes in the currents or patterns of flow of the animal spirits which the ventricles contain and which always tend to escape. Later we shall hear that the spirits flow out of the pineal gland especially freely toward pores that are relatively open and that the soul is able to sense the differential flow. See notes 72 and 133, and Descartes's *Principles*, AT 8:319–323 (AT 9:314–317).

essential that its place of residence be preserved) to experience a feeling of *pain*.[66] [Cl–30; AT–144]

And if they are pulled by a force almost as great as the preceding without, however, being broken or separated from the parts to which they are attached, they will cause a movement in the brain which, testifying to the good constitution of the other parts, will cause the soul to feel a certain corporeal sensual pleasure referred to as tingling,[67] which as you see, being very close to pain in its cause, is quite the opposite in effect.

If many of these filaments are pulled equally and all together, they will make the soul sense that the surface of the object touching the member where they terminate is smooth; and if they are pulled

66. All the major physiological systems of antiquity — Platonic, Aristotelian, Epicurean, Stoic, Galenic — included explanations of pain and pleasure. In the century preceding Descartes, little was added to Greek interpretations of these phenomena — little, at any rate, that appears to have influenced him. Even Plato and Aristotle were somewhat parsimonious on the physiological basis of pain and pleasure, and their ideas are interesting mainly as background for the more extensive treatment accorded the topic by Galen.

In the *Timaeus* (64b–65a, Cornford trans.), Plato depicted sensation as a propagated dislocation of body particles, "one particle passing on the same effect to the other, until they reach the consciousness," while pain occurs when "the disturbance of the normal state is sudden, and the restoration gradual and difficult." Aristotle saw pleasure and pain, physiologically considered, as alterations of the mean condition of the body "toward what is good or bad as such" and hence to be desired or avoided (*de An.*, 431a10–19). Galen was influenced by these and other pre-Galenic ideas but carried the analysis further than any of his predecessors. Moreover, it was Galen's formulation that led to that proposed by Descartes.

Of particular import was Galen's belief that pain results from a rapid and violent change of constitution, or a "dissolution of continuity," occurring within the sense organ or other painful tissue (*Ars med.*, chap. 20, K1, 357; *de Loc. aff.*, bk. 2, chap. 5, K8, 80; *Meth. med.*, bk. 12, chap. 7, K10, 852; and especially *Sympt.*, bk. 1, chap. 6, K7, 115–117). Pleasure Galen regarded as an appreciably rapid restoration of the painful disruption.

Descartes differs from Galen on the physical basis of pain in specifying the occurrence of a break not in the organ where the painful sensation arises but in the fibrils of that organ's nerve or nerves. This leads to actual detachment of the fibrils from the organ in question. Elsewhere, Descartes distinguishes pain itself from the externally visible movements that betoken its presence. He says that in man both pain and the concomitant visible movements are found; in animals, the visible movements but not the pain itself. This squares with his idea that pain is a passion — a feeling — of the soul, and his insistence that in animals no soul exists (letter to Mersenne AT 3:85). For sensations disagreeable without being painful, see note 82 and Rothschuh, *D.*, 71, note 2.

67. Descartes uses the word *chatoüillement*. The suggestion is that slightly excessive but noninjurious stimuli can be agreeable if the constitution of the body is sound. See also the *Principles*, AT 8:318 (9:313), and especially the *Passions*, AT 11:398–399.

unequally, they will cause the soul to feel that it is uneven and rough.

If they [the nerve filaments] are set in motion only slightly, and separately from one another, as they continually are by the heat that the heart communicates to other members, the soul will have no more sensation of this than of all other ordinary actions;[68] but if this movement is augmented or diminished by some unusual cause, its augmentation will make the soul have a feeling of heat; its diminution, a feeling of cold.[69] And finally, according to the divers other ways in which they are moved, they will cause [the soul] to sense all the other qualities which belong to touch in

68. By "ordinary actions" Descartes probably means actions (visceral or muscular) of which we have no conscious sensation.

69. *Hot and cold sensations*. To explain these Descartes made use of Greek ideas modified in unimportant ways by pre-Cartesian Renaissance thinkers and adapted by him to his own corpuscular concept of the structure of the body. We look then to ancient rather than to medieval or Renaissance sources for Descartes's ideas on this subject.

Plato, for example, had ascribed the hot sensation to the "rending and cutting effect of [sharp-edged tetrahedral] fire [particles] upon our body." He seemed to see the sense of cold as arising from an undue compaction and immobilization of body particles, although the details are neither simple nor clear as he presents them. For the complexities of his theory of hot-cold sensitivity, see F. M. Cornford, *Plato's Cosmology*, London, 1937, 260–262.

Aristotle's theory of sensation omitted corpuscular considerations and posited instead a mean condition of the body with respect to polar opposites — in the present case, hot and cold. The cardinal feature of his doctrine is that the intermediate or mean condition of the body is *potentially* what the object is *actually*. Before sensation occurs, he says, the sense (that is, the body's mean condition) is "potentially either [hot or cold] but actually neither," and sensation is a process in which the body's *potentiality* to be like the object [that is, to be hot or cold] is *actualized*. Expressed differently, the body momentarily acquires the form of the object (*de An.*, 423b26–424a10).

Galen thought the balance of polar opposites, hot-cold and dry-moist, was perfectly tempered in the skin of the human hand; he supposed the hand to be exactly intermediate — in its temperament or qualitative blend — between the extremes of boiling water and numbing ice. This permitted the hand to serve as a standard of comparison for estimating heat and cold in objects touched (*Temp.*, bk. 1, chap. 8, K1, 559 and bk. 2, chap. 1, K1, 572).

Descartes's theory may be regarded as a mechanized version of Galen's adaptation of Aristotle's notion that sensation involves departure from a mean. In Descartes's theory, heat sensations are initiated when environmental particle motion (heat) causes particle motion (heat) in the skin or some other heat-sensitive part of the body (*Light*, AT 11:10). That the skin responds by particle motion is evidenced by the fact that rubbing (friction) too can warm both living and nonliving things. The theory differs neurologically from Galen's, which distinguished sensory from motor nerves. I have not discovered that Descartes specifically analyzed hot and cold in terms of different activities of the nerve, nor that medieval or early Renaissance innovations, or variations on the Galenic scheme, influenced him to a significant degree.

general, such as humidity, dryness, weight, and the like. [Cl–31; AT–145]

However thin and mobile these filaments may be, they are still not mobile enough to report very subtle natural events to the brain. The subtlest events they report are ones engaged in by the coarsest particles of earthy bodies.[70] Indeed there may even be some bodies whose parts, even though coarse, will slide against the filaments so lightly that, even though they press against or even cut completely through them, their action fails to pass to the brain — just as there are certain drugs that have the power of inactivating or even corrupting the parts to which they are applied without causing us to have any sensation.[71]

But the filaments that compose the marrow of the nerves of the tongue and that serve as an organ for *taste* in this machine, can be moved by lesser actions[72] than those that serve for touch in general,

70. Descartes means that rather coarse and gross "actions" (corresponding approximately to what we think of as stimuli) are needed to produce even a minimal effect on the nerve.

71. Descartes seems to wish to make room in his theory for subliminal nervous activity, which he interprets as activity that disturbs the nerve without causing any corresponding disturbance in the brain.

72. *The common sense.* How to distinguish among the modes of sensation — sight, hearing, smell, taste, and touch? Aristotle had postulated a sixth or "common" sense that differentiated among the other five (*de An.*, 426b8–427a15). He also tentatively associated the different receptivities of the sense organs with differencs in their constitution (sight with water, sound with air, smell with fire, taste and touch with earth; *de An.*, 435a–b). Galen was more specific; he ascribed vision to a luminous, hearing to an aerial, smell to a breath-like, taste to a watery, and touch to a solid constitution of the organ in question (see, for example, *Sympt.*, bk. 1, chap. 6, K7, 122).

In pre-Cartesian Renaissance thought, the idea of a *sensus comunis* was generally retained (see Fernel, *Physiologie*, 428–431; Piccolhomini, *Praelectiones*, 269; and du Laurens, *Historia*, 392). Its location (the *sensorium communum*) was a subject of debate (as evident from Vesalius, *Fabrica*, 623; Fernel, *Physiologie*, bk. 5, chap. 8; Bauhin, *Theatrum*, 569; and Bartholin, *Institutiones*, 212).

Descartes himself considers elsewhere three attributes of the common sense: its location (he places it in the conarium; letter to Mersenne, AT 3:361 and *Dioptrics*, AT 6:129); its ontological status (he views it as corporeal, as a physical body occupying space); and its role in conscious perception. He bases discrimination on a series of changes — a causal rather than a temporal series, since the changes are simultaneous. The causal chain comprises the following events. First, external action produces a spatial reconfiguration of the outer sensory membrane, much as sealing wax is reconfigured by a seal. Second, through nerve action, a corresponding — but not identical — change occurs in the common sense, whose total configuration at each moment registers the changes occurring in sensory membranes wherever they exist in the body. Third, the common sense "impresses" (spatially reconfigures) the imagination which, like the common sense, is a body occupying space (*res extensa*). The

both because they [the former filaments] are a little finer [than the latter] and because the membranes that cover [the former] are more tender.

Assume, for example, that they [the filaments] can be moved in four different ways by part[icle]s of salt, acid, common water, and *eaux de vie*, whose shapes and sizes I have already explained,[73] and that thus they can cause the soul to sense four different kinds of tastes. This they can do inasmuch as salt particles, being separated from one another and agitated by the action of the saliva, enter, pointed end foremost and without bending, into the pores in the membrane of the tongue. Acid particles flow on the diagonal, slicing or cutting the tenderest of [the tongue's] parts while yielding to the coarser. Those of fresh water merely glide past on top without incising any of its parts or advancing very far into the pores. Finally, those of *eau de vie*, being very small, penetrate the most readily of all and are moved with very great speed. Whence it is easy for you to judge how the soul will be able to sense all the other sorts of tastes — if you consider in how many other ways the particles of earthy bodies can act against the tongue.[74] [Cl–32; AT–146]

---

configurations existing in the imagination at each moment are designated "ideas," which thus have a physical or material nature. Fourth and finally, the impressed imagination in turn impresses "that by means of which we know," namely incorporeal awareness (*res cogitans*, or soul). We hear also that the soul can receive impressions from the common sense directly, without the intermediation of the imagination.

For these and related details of Descartes's theory of sensation, see his *Regulae*, AT 10:413–415, and Gilson's references to Descartes on the *sens commun*, ISC, 263; also notes 62 and 133. The wax analogy occurs in Aristotle, *de An.*, 624a17–23, and often in medieval and pre-Cartesian Renaissance authors.

73. See the *Meteors*, AT 6:233–238, and letter to Plemp, AT 1:422–424.

74. *The sense of taste*. Plato and Epicurus ascribed differences in taste to differences in the shape, size, and motion of the particles of the tasted substance (*Timaeus*, 65c–66b; Lucretius, *On the Nature of Things*, bk. 2, lines 398–407). For Aristotle, taste entailed an actualization of the tongue's potentiality for acquiring the form of the tasted object, a form tending toward one or the other of the two pairs of opposites, sweet vs bitter and oily vs saline, or toward certain intermediates — acid, pungent, astringent, sharp (*de An.*, 422a–b). Galen weighed Aristotelian and other ideas about primary tastes, and himself analyzed tastes in terms of (*a*) their reducibility to the four primary qualities (hot, cold, dry, moist) and (*b*) their physical modes of action (whether astringent, compressive, penetrant, incisive, cleansing, and the like); he added that the tongue judges not only in terms of the foregoing criteria but also — as we are about to hear — in terms of assimilability (see note 75). For Galen on taste, see *de Simp. med.*, bk. 1, chaps. 38, K11, 450–455 and bk. 4, chaps. 8 and 9, K11, 647–651; also *de Plac.*, bk. 7, K5, 627–634.

But what must be chiefly noted here is that the particles of food which, while still in the mouth, can enter the pores of the tongue and elicit the sense of taste are the same ones which, once in the stomach, can pass into the blood and proceed thence to join and unite with all the parts of the body. And indeed only those that moderately tickle the tongue, thereby causing the soul to sense an agreeable taste, will be entirely suitable to the end [of incorporation into the body].[75]

For, just as particles that are too active or too inactive can cause too sharp or too bland a taste, so are they too penetrant or too soft to enter into the composition of the blood and serve for the maintenance of the members. Moreover, there are some particles that cannot give the soul a sensation of taste or savor and hence are generally unfit to be swallowed. These include: [a] particles that are so coarse or so strongly joined to one another that they cannot be separated by the action of saliva; [b] particles that cannot penetrate the pores of the tongue so as to act on the nerve filaments that serve there for taste (except as they might act on those that serve, in other members, for the general sense of touch); and [c] particles lacking pores in themselves into which the particles of the tongue,

---

Vesalius and post-Vesalian physiologists (Fernel, Columbus, Piccolhomini, du Laurens) are remarkable for the scant attention they gave to the sense of taste. Exceptions may be mentioned (for example, Bauhin, *Theatrum*, 982–987 and Crooke, *Mikrokosmographia*, 714–726), but such accounts contained little that was innovative or that appears to have inspired the particle-based theory of Descartes. There is a note on the sense of taste in the *Principles*, AT 8:318 (9:313).

75. Descartes is theorizing here on the relation of taste to assimilability, a field extensively worked since Greek times. It had long been assumed that the taste of the food is somehow connected with its suitability or salubriousness. The problem: to explain the connection. The question was already current when the author of the Hippocratic treatise *Ancient Medicine* (circa 400 B.C.) speculated that medicine began when men learned to avoid the crude diet they had formerly shared with the beasts (see chapter 3). Galen argued often in the vein that "soft and friable food is maximally conducive to health" (*de Alim. fac.*, bk. 3, chap. 30, K6, 726), and that in healthy persons the foods that taste most agreeable are the ones most readily assimilated (*de San.*, bk. 6, chap. 3, K6, 394).

Descartes was probably familiar with Fernel's crtical treatment of something called the "natural appetite." Through this appetite, according to certain thinkers, men and animals consider pleasant — and their body parts attract — what is similar, proper, and benign. Even in plants, some have supposed that the intimate parts have a natural appetite for substances similar and advantageous to themselves (*Physiologie*, bk. 5, chap. 5, 412–417). Whatever his immediate inspiration, Descartes reformulates the problem in his usual "corpuscularizing" manner (see also the paragraphs immediately following).

or at least those of the saliva with which it is moistened, might enter. [Cl–33; AT–147]

And the foregoing point is so generally true that often, in the measure that the stomach's temperament changes, the strength of taste changes also; so that a food that usually seems agreeable in taste to the soul may at special times seem bland — or bitter: the reason for which is that the saliva, which comes from the stomach and retains the qualities of the humor that abounds there, is mixed with food particles in the mouth and contributes much to their action.

The sense of smell,[76] also, depends on many filaments which advance toward the nose from the base of the brain below those two

76. *Olfaction.* Descartes builds his theory on a tradition traceable to Galen. The latter had regarded olfaction as the only sense mediated entirely within the brain itself — the other senses (sight, hearing, taste) involving, in addition, certain more or less nerve-like protrusions of the brain substance. These protrusions were thought to extend the brain's sensory faculties to the sense organs. Odors, Galen reasoned, must pass through apertures in the ethmoid bone and through the presumably permeable floor of the brain ventricle itself. The same apertures were supposed to permit the inflow of air into the brain for conversion into psychic pneuma as well as an outflow of fluid excesses (when acute, otherwise they drained off by way of the palate; see note 80 and Galen's *UP*, bk. 8, chaps. 6–7, K3, 647–656). Galen complained that "a majority" considered the olfactory organ to lie beyond the cranial cavity in the nasal passages (*de Plac.*, bk. 7, chap. 5, K5, 628).

During the sixteenth century the status of the "mammillary processes" (our olfactory tracts) was debated with special reference to whether they were or were not nerves. Vesalius remained noncommittal on the point (*Fabrica*, 322–323 and 643). Niccolo Massa supposed that beneath each of the two terminal thickenings (our olfactory bulbs) a nerve descended through the ethmoid bone to the nostrils, there to act as the organ of smell (*Epistolae medicinales*, Venice, 1550, 58 and *Anatomiae liber introductorius*, Venice, 1559, 87). Columbus seems to have regarded the terminal thickenings themselves as the sense organs of smell (*Anatomica*, 193–194). Piccolhomini reasoned similarly, and called the strands connecting these thickenings with the brain posteriorly (our olfactory tracts) olfactory nerves (*nervi odorati*) (*Praelectiones*, 292). Bartholin took the same position (*Institutiones*, 1611, 399), positing the existence of olfactory nerves (where we see the olfactory tracts of the brain).

Descartes correctly locates receptor action in fibrillar derivatives of the tips of the mammillary processes (olfactory bulbs); however, he incorrectly sees these as extending not all the way to the nasal membrane, but only as far as the subdural space (between the pial and dural meninges of the brain itself). Whence Descartes's idea of olfactory fibers? From dissection? From earlier authors? Through logical deduction? The question remains unsettled. What we know to be olfactory fibers had already been described by Vesalius, and by later anatomists, but were not assigned an olfactory function and were wrongly derived from (what these anatomists believed to be) the third cranial nerve.

That olfactory fibrils traverse the cribriform plate was urged after Descartes's death by Conrad Schneider (*De osse cribriformi . . .*, Wittenberg, 1654), but Schneider continued the error of deriving these from the "third" cerebral nerve. Ten years later

little hollowed-out parts that anatomists have compared to nipples.[77] These [filaments] differ in no way from the nerves that serve for touch and taste except that [a] they do not leave the cavity of the head which contains the brain as a whole; and [b] they can be moved by smaller earthy part[icle]s[78] than can the nerves of the tongue, both because they are slightly finer and because they are more immediately touched by the objects that move them.[79] [Cl–34; AT–148]

For you should know that when this machine breathes, the subtlest air part[icle]s that enter it through the nose penetrate, through pores in the bone termed spongy [the ethmoid bone], if not all the way to the cavities of the brain then at least to the space between the two membranes that envelop it whence these particles can go out again through the palate; just as, reciprocally, when air leaves the chest it can enter this space through the palate and leave again by way of the nose.[80] [You should know] too that

---

Thomas Willis rightly understood the olfactory nerves to be the first cerebral pair with terminal fibers traversing the cribriform plate to "serve properly for the very Organ of Smell" (*Cerebri anatome . . .* [first publ. 1664], trans. S. Pordage; republished in *Dr. Willis's Practice of Physick*, London, 1684, 112). There is a note on the sense of smell in Descartes's *Principles*, AT 8:318–319 (9:313).

77. The word *mammelles is* used. Noting the absence of conspicuous terminal swellings in human as opposed to animal olfactory tracts, Vesalius had distinguished "what the vulgar term mamillary bodies" (he made these integral parts of the anterior tips of the cerebral hemispheres) from "organs of olfaction" (our olfactory tracts, which Vesalius represented as if they lack terminal bulbs altogether) (*Fabrica*, 318–323). The "mammelles" which Descartes evoked were probably olfactory bulbs that he saw in the brains of animals and, not entirely incorrectly, assumed to be present in man (in man, they are less conspicuously developed).

78. In other words, particles of the third element that impinge on the nerve endings and evoke the sense of smell.

79. In the case of other sensations, longer nerves intermediate between the sense organs and the brain.

80. Descartes's wrong idea of a connection between the intermembranal (subdural) space and the pharynx derives indirectly from Galen, who thought excess and excremental brain fluid ordinarily drained from the ventricles through the infundibulum to the palate (through openings in the sphenoid bone) but, when acute, to the nostrils (through pores in the cribriform plate). The relevant passages in Galen are UP, bk. 8, chap. 6, K3, 650; and *de Plac.*, bk. 7, chap. 3, K5, 609.

Galen's idea continued the Aristotelian belief that nasal secretions derive from the brain (*PA*, 653b36–655a3). Rothschuh, D., 75, note 3, reminds us that many physiologists, among them Fernel, advocated a conception similar to that of Descartes, namely that through a pulsating movement of the brain, air is drawn in between the brain membranes (Fernel, *Physiologie*, bk. 5); but this had also been denied — for

on entering this space [between the brain envelopes] the [air particles] encounter the extremities of [olfactory] filaments [and that these are] quite bare or covered only by a membrane which is so extremely delicate that little force is needed to move them.

You should know also that these pores [in the ethmoid bone] are so arranged, and so narrow, that they admit to these filaments no earthy particles coarser than those which, when speaking earlier on this subject, I designated "odors;" except, perhaps, for certain ones that constitute *eaux de vie*, because the shape of these renders them strongly penetrant.[81]

Finally, you should know that among those extremely small earthy particles that always exist in greater abundance in air than in any other composite bodies, only those which are a little coarser or finer than others, or which because of their shape are more or less easily moved, can cause the soul to sense a variety of odors. And indeed only those in which these extremes are much moderated and mutually tempered will make it have agreeable [sensations]. For those [particles] that are only normally active will not be perceptible at all; and those that act with too much or too little force will necessarily be unpleasant.[82] [Cl–35; AT–149]

As to the filaments that serve as a sense organ of hearing,[83] they need not be as thin as the preceding. It suffices instead to suppose:

---

instance by Riolan ("Anthropographie," *Oeuvres*, 1629). Du Laurens devoted a chapter to the clearance of cerebral excesses (*Oeuvres*, 1621, 321–322) detailing the above, and other, ideas of Galen, and summarizing sixteenth-century treatments of the subject. After the time of Descartes, Conrad Schneider argued against the drainage of the brain pan (*Liber de osse cribriformi*, 1655, 69–75). For an account of further developments, see G. Prochaska, *Functions of the Nervous System* (1784), London, Sydenham Society, 1851, 363–380.

81. The particles of odorous substances must be very fine to penetrate the tiny channels of the ethmoid bone. Note that, contrary to prevailing (neo-Galenic) ideas, air particles are excluded. Galen and his medieval and Renaissance followers permitted air to pass through and become converted to animal spirits.

82. The point that Descartes wishes to make about pleasant and unpleasant odors is that ordinary activity elicits no conscious response. Moderately increased or decreased activity is agreeable; immoderately increased activity, disagreeable. See also note 67.

83. *Hearing*. Most earlier theories of hearing had made much of the supposition that the sensitive part of the ear contains either air or a substance similar to air. Ancient doctrine on the ear had reached a kind of culmination with Galen whose theories were continued, with rather extensive modifications, by medieval and

[a] that they are so arranged at the back of the ear cavities that they can be easily moved, together and in the same manner, by the little blows with which the outside air pushes a certain very thin membrane [the tympanum] stretched at the entrance to these cavities; and [b]that they [these filaments] cannot be touched by any other object than by the air that is under this membrane. For it will be these litle blows which, passing to the brain through the intermediation of these nerves,[84] will cause the soul to conceive the idea of sound.

Renaissance medical thinkers. Aristotle had postulated an internal air continuous with external air and responsible for hearing (de An., 420a3–420b5). This idea Galen fitted into his more general concept that hearing is airy, smell vaporous, taste watery, and touch earthy. For Galen the air in the labyrinth of the ear was the primary organ of hearing. He also said that the labyrinth of the ear is reached by a prolongation of the brain (that is, the auditory nerve). UP, bk. 8, chap. 6, K3, 645–646; also de Plac., bk. 7, chap. 5, K5, 627.

In the mid-sixteenth century, Fernel retained the Galenic idea of "a certain subtle air placed in the ears at birth, enveloped in a membrane, and situated at the back of the ears to which come the auditory nerves that rise in the brain" (Physiologie, bk. 5, chap. 7). In the post-Vesalian period, the anatomy of the ossicles of the ear became known primarily through the work of Ingrassia (1544) and Eustachius (1563), but progress beyond a Galenic view of hearing was slow. In 1586, Piccolhomini retained Galen's view of an innate aerial sensorium of hearing within the ear, its contents derived from the purest components of paternal semen and maternal blood. The aerial substance, he said, is enclosed by a membranous enlargement of the meninges of the auditory nerve, and the ear ossicles imprint it with the species — in Peripatetic language, the form — of the sound (Praelectiones, 298–301).

In 1600, in a treatise devoted to hearing, Fabricius postulated as fundamental to hearing an intimate commixture of (a) the innate air with (b) animal spirits bringing with them the auditory sentient faculty of the brain (De aure, auditus organo . . ., publ. with his De visione and De voce, Venice, 14–15 and 34–36). In the same year, however, du Laurens argued against the innate air as auditory organ on the basis that this air is inanimate and subject to continuous renewal. He wished to make the auditory nerve the sound organ just as the mammillary processes are organs of smell (Historia, 428–429). Bauhin (Theatrum, 802–858) argued not much differently, making the ear ossicles help to communicate the "sounds or their characters" from the external medium (the atmosphere) in the form of movements to the internal medium (the air enclosed in the labyrinth) which in turn communicates sounds or characters to the auditory nerve. A kind of synthesis of Fabrician and Laurentian ideas was espoused by Riolan (1618), who made the resonating included air actually enter the auditory nerve endings and mix with animal spirits (see "Anthropographie," Oeuvres, 649).

What was needed was an outright rejection of Galenic faculty-physiology. Descartes took this necessary step and deserves the credit for so doing, despite the complete inadequacy of the model of afferent transmission that he substituted. A note on hearing is included in his Principles, AT 8:319 (9:314).

84. In this passage on conduction, we are told that "blows" (secousses) "move" (passent) all the way to the brain, but Descartes did not believe that any sort of change took time in traveling along the nerve. The whole nerve responds simultane-

Note that a single such blow will be able to cause nothing but a dull noise which ceases in a moment and which will vary only in being more or less loud according as the ear is struck with more or less force. But when many [blows] succeed one another, as one sees in the vibrations of strings and of bells when they ring, then these little blows will compose one sound which [a] the soul will judge [to be] smoother or rougher according as the blows are more or less equal to one another, and which [b] it will judge [to be] higher or lower according as they succeed one another more promptly or tardily, so that if they are a half or a third or a fourth or a fifth more prompt in following one another, they will compose a sound which the soul will judge to be higher by an octave, a fifth, a fourth, or perhaps a major third, and so on. And finally, several sounds mixed together will be harmonious or discordant according as more or less orderly relations exist [among them] and according as more or less equal intervals occur between the little blows that compose them.[85] [Cl–36; AT–150]

---

ously, just as the top of a quill pen moves when one writes with the point (see notes 45 and 58). In the *Sixth Meditation*, he makes the crucial point that sensory nerves produce the same sensation whether stimulated at the extremity in the normal way or at some point intermediate between the sense organ and the brain (AT 7:88).

85. Descartes's musical theories, like his theories on most of the subjects treated in *Man*, were adaptive modifications of borrowed ideas. They were interpretations, in terms of Cartesian particle theory, of notions derived from earlier Renaissance authors and from Descartes's own contemporaries. A classic model for the alliance of particle theory with the mathematics of music had been supplied in antiquity by Plato. In his *Timaeus* (67a–c), he defined sound as the impact created by air that had gained entrance to the blood and the brain; he supposed that this impact gave rise to motions that were transmitted through the body until they arrived at the soul. "A rapid motion produces a high sound," Plato said, and he added that "the slower the motion, the lower the sound." Moreover, "if the motion is regular, the sound is uniform and smooth; if not so, harsh. Depending whether the movement [of the particles of the body responding to the impact of particles of air] is of great or small magnitude, the sound is loud or soft." Plato had, in addition to these ideas, a theory of agreeable concordance (see F. M. Cornford, *Plato's Cosmology*, 320–326).

In Descartes's *Compendium musicae* and elsewhere, he mentions — and he plainly owes much to — Gioseffo Zarlino, whose *Institutione harmoniche* appeared in Venice in 1558. The decades between Zarlino and Descartes saw an outpouring of books by many authors devoted wholly or partialy to musical theory, and Descartes's indebtedness to them has been analyzed by André Pirro, *Descartes et la musique*, Paris, 1907.

Another influence on Descartes's theory of music was that of his constant correspondent Marin Mersenne, with whom he exchanged ideas about musical problems at approximately the time when *Man* was composed. Professor Crombie reminds me that Mersenne wrote about vibrating strings and about consonance and dissonance in works published between 1623 and 1627, works likely to have been familiar to Descartes.

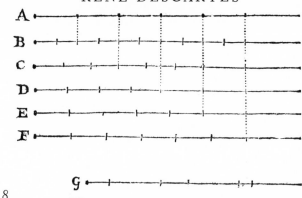

Figure 8

As for example [Fig. 8], if the divisions of the lines A, B, C, D, E, F, G, H represent little blows that constitute as many different sounds, it is easy to judge that those that are represented by the lines G and H cannot be as smooth to the ear as the others: [just] as pieces of stone joined end to end are not as smooth to the touch as those of a well-polished mirror. And one must think that B represents a sharper [higher pitched] sound than A by an octave, C by a fifth, D [by] a fourth, E by a major third, and F by a major full tone. And [one must] note that A and B joined together [the octave do:do], or ABC [do:sol:do], or ABD [do:fa:do], or even ABCE [do:me:sol:do], are more concordant than are A and F [do:re], or ACD [do:fa:sol], or ADE [do:me:fa], and the like. This seems to me sufficient to show how the soul which will be in the machine that I am describing will be able to enjoy a kind of music following all the same rules as ours; and how [the soul] will even be able to perfect [that music] — at least if one considers that it is not absolutely the smoothest things that are most agreeable to the senses, but those that titillate them in the best-tempered way: just as salt and vinegar are often more agreeable to the tongue than fresh water. And it is this that makes music as receptive to thirds and sixths and even sometimes to dissonances as to unisons, octaves, and fifths.[86] [Cl–37; AT–151]

86. The bracketed items in the text translate the various consonances and dissonances into chords. See Descartes's *Compendium of Music* for more detail on consonances (AT 10:96–111) and dissonances (AT 10:127–131). Near the beginning of the *Compendium*, Descartes says that "those proportions [in duration and in pitch] are most agreeable in which the relations are neither too easy nor yet too difficult to perceive" (AT 10:92). When he says here that the machine's soul would, like man's,

There still remains the sense of *vision*, which I need to explain a little more exactly than the other senses because it is more useful to my subject. This sense depends, in this machine [as in us], on two nerves which must doubtless be composed of many filaments. These filaments must be as delicate and as easily movable as possible, inasmuch as they are destined to report to the brain the divers actions of the particles of the second element — which actions, in accordance with what has been said earlier, will enable the soul, when united with this machine, to conceive the divers ideas of colors and of light.[87] [Cl–38; AT–151]

But because the structure of the eye also aids in [producing] this result, it is necessary that I describe it, and for greater ease I shall try to do so in few words, intentionally leaving out many superfluous details which the curiosity of the anatomists has uncovered there.

------

be able to render the music more perfect, he probably means that basically agreeable consonances can be made even more agreeable by the addition of extra notes (for example, the insertion of a fifth within an octave) or by moving from one agreeable consonance to another, "as we become disgusted sooner from eating only sugar or other similar dainties than from eating only bread which everyone affirms to be less pleasing to the palate" (AT 10:106).

87. *Light*. This topic is considered by Descartes in the treatise of that name and in the *Dioptrics*. The presentations differ in emphasis but are consistent in their essentials. The agreement between them is to be expected, since they were written at about the same time. The treatment in the *Treatise of Light* is unattractively deductive and complex. There is some question as to whether it can really be understood at all, but the object is to show how second-element particles circling about a light source tend to generate a radial, centrifugal motive tendency; it is this tendency, this unexpressed disposition or effort of second-element particles to move radially, that constitutes light.

Intrinsically, the revolving second-element particles tend to move tangentially, Descartes believes, but they are kept in orbit by the restraint of particles farther out (somewhat as a stone being whirled in a sling is restrained from flying off by the cord).

If, from the light source considered as a base, a cone is erected, the centrifugal tendencies of particles within the cone will impinge on its apical particle, inclining it to move straight away from the source. There is no actual outward movement of the apical particle, however, merely the communication of an outward-moving tendency, or effort, along the ray considered as axis of the cone. In vision, this effort (light) is communicated to the eye.

The transmission of the radial tendency or effort of second-element particles is abetted by movements of the first-element particles of which the light source consists. These move out centrifugally, reinforcing the outward tendency of the second-element particles. In the case of the first-element particles, their outward or centrifugal movement is compensated by a simultaneous contripetal movement — in one of those cycles of displacement and replacement which Descartes likes to imagine. For these ideas, see AT 11:84–97. Descartes regards his model as equally adequate to explain photoemanation from a star and from a glowworm (letter to Mersenne, AT 2:179–180).

[In Fig. 9] *ABC* is a rather hard and thick membrane [the sclera] that constitutes a round vase, as it were, in which all the other parts of the eye are contained. *DEF* is another, thinner membrane [the choroid], spread like a tapestry inside the preceding. *GHI* is the nerve [the retina] whose little threads (*HG, HI*) — spreading in all directions from *H* as far as *G* and *I* — entirely cover the back of the eye. *K, L,* and *M* [the aqueous humor, lens, and vitreous humor] are three sorts of extremely clear and transparent albumins or humors which fill all the space in the interior of these membranes and which have, respectively, the shapes pictured here.[88] [Cl–39; AT–152]

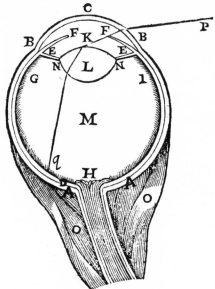

Figure 9

88. *The eye.* On the eye as on many physiological subjects Descartes borrowed — but also radically altered — the ideas of earlier authors. Galen's terminology had included the words "sclera," "choroid," "aqueous humor," "crystalline humor," and "vitreous humor," his crystalline humor being our lens. Galen considered the retina to be an expansion of the optic nerve and hence an extension of the brain. He viewed the lens as a receptor, not a refractor. The retina perceives alterations producd by light in the lens, Galen said, being able to do this because the endings of the expanded optic nerve (the retina) are actually inserted into the lens around its edges. He viewed the choroid as an extension of the pia mater of the brain, and the sclera as an extension of the dura mater (all cranial nerves being covered by such membranes). He described the intimate relations of the tunics and humors in detail, but not without certain errors (UP, bk. 10, chaps. 1–6, K3, 759–790).

In the first membrane, the part *BCB* [the cornea] is transparent, and a little more arched than the rest; and the refraction of rays entering it occurs toward the perpendicular. In the second membrane [the iris], the internal surface of the part *EF* [our pupillary sphincter and dilatator muscles], which faces the back of the eye, is completely black and opaque,[89] and has at its center a little round hole that is called the pupil and appears black at the middle of the eye when one looks at it from without. The hole is not always of the same size, because part *EF* of the membrane that the hole is in, swimming freely in humor *K*, which is very liquid, seems to be like a litle muscle that is enlarged or diminished under the direction of the brain as use requires.

The shape of the humor marked *L*, which is called the crystalline humor [the lens],[90] is like the shape of the glasses I described in the

---

During the pre-Cartesian Renaissance, a majority of medical and physiological theorists accepted the main tenet of the Galenic theory — namely, that animal spirits in the optic nerve and retina extend to the lens the "visive" faculty of the soul — while arguing about such questions as the number of the eyes' tunics, whether the optic nerves were hollow, and what precisely were the roles played by the aqueous and vitreous humors. What was needed was a radically new view of the eye in which the lens serves not as a receptor but as a refractor that projects an image onto the true receptor, the retina, at the back of the eye. The events that led to this important step are outlined in note 90.

89. The pigment of the iris is actually dark purple rather than black and is limited to the inward-facing surface in blue-eyed persons; in others, it is present also in the overlying, outer cell layers. For more of Descartes's views on this subject, see the *Dioptrics* (AT 6:107). A little earlier, Christopher Scheiner had discussed the eye pigments in his *Oculus, hoc est fundamentum opticum* . . ., Innsbruck, 1619, 29–30. Descartes referred to Scheiner occasionally in his letters and probably knew his published works. This was the same Scheiner who argued with Galileo about sun spots and helped to secure the latter's condemnation by the Inquisition.

90. *Physiological optics.* This science as we understand it could not have emerged had not the Galenic idea of the lens as a sensor given way to the proper view of this structure as a refractor. The transition from the old view to the new one was gradual, commencing in late Arab times and culminating in the sixteenth century. As the role of the lens as refractor began to be appreciated, its role as sensor was not immediately rejected. For example, Alhazen (d. circa A.D. 1035) assigned it both roles. That sensation should be moved back from lens to retina did not occur to him. Although he acknowledged the lens's refractive function, he failed to grasp that refraction leads, in the eye, to image formation. He did not think, as far as the eye was concerned, in terms of a two-dimensional pattern projected onto a surface or screen a certain distance behind the lens. He thought the rays — or perhaps their species or form — on leaving the back of the lens were re-refracted and made to diverge so that they would not meet and cross and thus undergo an inversion. Curiously, despite increasing realization that lenses are producers of images, the transfer of the sensorial role from lens to retina was consummated only at the end of the sixteenth century.

treatise on *Dioptrics*,[91] by means of which all the rays that come
from certain points are reassembled at certain other points; and its

---

The interim history of these developments is too complex to be treated here, even
in outline. What we chiefly need, in order to understand the theories of Descartes, is
information about the interpretation of the lens in the period immediately preceding
his own.

Vesalius pleaded ignorance with respect to the uses of the parts of the eye (*Fabrica*,
649–650). Columbus referred to the crystalline humor as *permaximus*, as *tollendus*,
and mistakenly viewed it as "the chief organ of vision"; but he correctly located it
forward of the center of the eye (an improvement over earlier authors, including
Vesalius, who located it at the center), and he noted that, assuming it to be movable,
one might obtain both magnification and resolution as in the case of the surgical
speculum (*Anatomica*, 219).

Later in the sixteenth century, the eye was compared, not for the first time, to a
camera obscura by Giambattista de la Porta (*Magia naturalis*, 1589 and *De refrac-
tione*, 1593) and especially by Felix Plater, who drew an historic distinction in assign-
ing an exclusively optical role (refraction) to the lens and a quite separate biological
role (sensory reception) to the retina (*De corporis humani structura et usu*, Basel,
1583, 187).

Plater's correct idea was not immediately accepted; the idea of lens as sensor was
retained by Fabricius, for example (*Visione*, 51–54 and 96–104), and by du Laurens
(*Historia*, 424). But Plater's view was adopted by Kepler, who went on to give a
classic and essentially correct interpretation of image formation (see note 96 and
Kepler's *Dioptrics*, sections 57 ff.), and by Christopher Scheiner who built models of
the eye (*Oculus*, see note 89 above, 70–76 and 114–214). Scheiner removed the
opaque layers at the back of extirpated eyes and showed that it was on their retinas
that images were formed (*Rosa ursina*, 1626, 60, cited by Crombie, see below).

We shall discover that Descartes was extensively influenced by Kepler, adding to
Kepler's analysis the important further point that the human lens accommodates to
distance by changes in shape (see note 99). Descartes mentions Scheiner's experiments in the Fifth Discourse of his *Optics*, but does not there mention by whom the
experiments were performed. Crombie is undoubtedly correct in thinking that Descartes repeated them himself.

The above outline, which makes no pretense to completeness, shows that far from
developing his physiology of vision *de novo* from self-evident axioms or ideas, Descartes carried on where a line of predecessors, culminating in Plater, Kepler, and
Scheiner, had left off. It is sometimes stated that Scheiner anticipated Descartes with
respect to the ability of the lens to accommodate by changes in shape, but I have
been unable to confirm this.

In the foregoing sketch I have profited from but not followed strictly A. C. Crombie's detailed and indispensable account of medieval and Renaissance optics (see his
"The mechanistic hypothesis and the scientific study of vision . . .," in S. Bradbury
and G. L'E. Turner, *Historical Aspects of Microscopy*, Cambridge, 1967, 3–112; also
his "Kepler: De modo visionis," *Mélanges Alexandre Koyré*, Paris, 1964, 139; and
two works of S. L. Polyak, *The Retina*, Chicago, 1941, 137–138, and *The Vertebrate
Visual System*, Chicago, 1957, 27–39). For further details of Descartes's theory of
the eye, see the Third to Sixth Discourses in his *Dioptrics*.

91. For these descriptions of optical glasses, see, in the *Dioptrics*, the Seventh Discourse ("Some means of perfecting the vision," AT 6:147–165), the Eighth
Discourse ("Shapes which transparent bodies must have in order to deflect rays
refractively in all the ways that serve vision," AT 6:165–196), and the Ninth Discourse ("Description of optical instruments," AT 6:196–211).

matter is less soft, is firmer, and consequently causes a greater refraction than that of the two other humors that surround it.

E and N are black filaments that come from within the membrane DEF and completely encircle the crystalline humor; they are like so many little tendons by means of which its shape can be changed and rendered a little flatter or a little more arched according to need.[92] Finally o,o are six or seven muscles[93] which are attached to the eye on the outside and can move it very easily and very quickly in all directions. [Cl–40; AT–153]

Now the membrane BCB [the cornea], and the three humors K, L, and M [aqueous, crystalline, and vitreous], being very clear and transparent, do not prevent the rays of light which enter through the pupil from penetrating to the back of the eye where the nerve is, nor from striking as easily against it as if it were completely exposed; they serve [rather] to protect it [the retina] against injuries from air and other external bodies that could wound it easily if they touched it; and [they serve] further to keep it so delicate and tender that there is no wonder it can be moved by acts so slightly perceptible as are those I here take to be *colors*.[94]

92. The "black filaments" probably refer to the processes of the ciliary body, which are covered posteriorly by pigmented layers. Vesalius referred to what we call the ciliary body as a "tunic arising from the iris" (*Fabrica*, 642).

93. Descartes here reflects the lack of agreement among his immediate predecessors and contemporaries as to the number of oculomotor muscles. Galen had sometimes erroneously extended to man the "seventh" (choanoid) eye muscle found in sheep and cattle and rudimentarily in certain primates (see, for example, UP, bk. 10, chap. 8, K3, 797). Vesalius failed to correct Galen on this point (*Fabrica*, 239–240). Columbus undertook a reclassification of the eye musculature which was unsuccessful but which recognized Galen's and Vesalius' error (*Anatomica*, 124). Fabricius, du Laurens, Bauhin, and Bartholin limited the eye muscles to six. For example, du Laurens said: "As to the seventh which almost all the anatomists describe, even Vesalius, and which they say surrounds the optic nerve and strengthens it to keep the eye from falling out of its orbit, this is found only in four-footed animals which always look down, and never in the human eye. The muscles of the eyes are, then, only six" (*Oeuvres*, 1621, 326–327). There is another mention by Descartes of the "six or seven muscles" in the *Dioptrics*, AT 6:108.

94. *Color*. In his *Dioptrics*, written before his *Treatise of Man*, Descartes treats color inadequately but makes two observations that will form the basis of his later, more definitive color theory. First, he says that when light is transmitted, second-element particles display two tendencies — one rotatory and the other rectilinear — comparable to the tendencies of a tennis ball when grazed, or cut. Second, he says that whereas the soul's cue to the intensity of light is the strength of the disturbance occurring in the brain at the point of origin of the optic nerve, its cue to color is the kind of disturbance that occurs there (AT 6:90–91 and 130–131).

The curve in that part of the first membrane marked *BCB* [the cornea], and the refraction that occurs there, enable rays from objects located toward the sides of the eye to enter through the pupil. They thus enable the soul, without the eye's moving, to see a larger number of objects than it otherwise could.[95] If, for example, ray *PBKq* [Fig. 9] did not curve at point *B*, it would be unable to pass [inside the circle marked by] the points *F,F* in order to arrive where the nerves are. [Cl–41; AT–154]

The refraction that occurs in the crystalline humor [the lens] serves to strengthen the vision and at the same time render it clearer. For you must know that the shape of this humor is accommodated to refractions occurring elsewhere in the eye and to the distances of different objects, so that when the vision is trained on a particular point of an object all the rays that come from this point, and that enter the eye through the pupil, are caused to reassemble at another point at the back of the eye. They assemble at precisely one part of the nerve that is there [the retina]; and, by the same means, other rays that enter the eye are prevented from touching the same part of this nerve.[96]

Only later, in the *Meteors*, in connection with his famous explanation of the rainbow, does he develop a definite corpuscular explanation of color. The color differences of light rays are determined, he believes, by the ratio of the rotatory tendencies of their second-element particles to the rectilineal tendencies of these particles. If the rotatory tendency or force is decidedly higher than the rectilineal, the color will be red; if only slightly higher, yellow. If the rotatory force is slightly less than the rectilineal, the color will be green; if decidedly less, blue; if still less, indigo (*incarnat*); and if even less than that, violet (AT 6:329–337). For further references to Descartes's discussions of color, see AT: "Index génerale," 76.

95. Descartes's point on extent of the peripheral field is factually true; how significant it is, adaptively, is difficult to say. Probably the chief adaptive role of corneal refraction is to bring a more central sector of the field into focus on the foveal region of the retina.

96. *Image formation.* Here Descartes continues the analysis, building especially on Kepler. In 1600 Fabricius, arguing incorrectly from the example of a vial filled with water and from the optics of eyeglasses, had postulated a series of refractions occurring when light passes successively through air, the cornea, the aqueous humor, and the lens. These refractions, he believed, cause light radiating from a point to reconverge to another point which he viewed, incorrectly, as lying within the light-sensitive lens (*Visions*, see note 90).

Fabricius' was the last notable attempt on the dioptric problem before its successful solution by Kepler (see his *Werke*, 2:162 and 4:368 and A. C. Crombie, *Kepler*, note 90 above, 147–162). Kepler acknowledged that the retinal image was inverted and reversed. That an inverted retinal image should be perceived as upright by the soul presented him with no special problem, since he departed completely from the medieval idea that the optic nerves convey light or the form of *species* of light. He

For example [Fig. 10], when the eye is so arranged as to look at point R, the arrangement of the crystalline humor makes all the rays RNS, RLS, and so forth, reassemble exactly at point S and, by the same means, prevents any of those that come from points T and X and so forth, from arriving there. It also assembles all those from point T in the neighborhood of point V, those from point X in the neighborhood of point Y, and so on. Whereas if no refraction occurred in this eye, object R would send only one of its rays to point S and the others would spread here and there throughout space VY; and similarly the points T and X, and all those in between, would each send its rays toward this same point S. [Cl–42; AT–155]

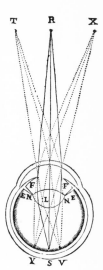

Figure 10

Now it is quite evident that if object R sends a large number of rays to point S, it must act more strongly against the part of the nerve that is there than if it sent only one; and that part S of the nerve must relay the action of object R to the brain more distinctly and faithfully if it receives rays from object R only, and not from divers others.

The black color both of the internal surface of the membrane EF [the iris] and of the filaments EN [the ciliary body] also serves to render vision more distinct: because, in line with what was said before concerning the nature of this color,[97] it deadens the force of rays reflected from the back of the eye toward the front and thus prevents them from returning toward the back where they might cause confusion. For example, if bodies [E]F [the iris] and [E]N [the ciliary body] were not black, rays from object X, arriving at point Y of the nerve (which is white) might be reflected thence in all directions toward N and F whence they could be turned back again toward S and V and there disturb the action of [rays that come from] points R and T.[98] [Cl–43; AT–155]

---

said that the information conveyed by nerves is, rather, a differentiation of animal spirits and quite outside the range of ordinary optical laws (see Crombie, 57–58).

Descartes made Kepler's concept of image formation more precise by submitting it to the new sine law of refraction first published by him, but developed earlier by Harriot and Snell. For Descartes elsewhere on the formation of the retinal image, see the Fifth Discourse of his *Dioptrics*.

97. Adam and Tannery found no previous allusion to the deadening quality of black color.

98. The eye does operate somewhat in the manner of a dark-chamber (*camera obscura*), whose dark interior helps, in the manner indicated by Descartes, to avoid internal reflective interference. As for the iris itself, its pigment renders it opaque, a condition without which (as in albinism) its effectiveness is reduced.

The change of shape that occurs in the crystalline humor permits objects at different distances to paint their images distinctly on the back of the eye. For, following what has already been said in the treatise on *Dioptrics*, if for example the humor *LN* [the lens — see Fig. 11] is of such a shape that it causes all the rays from point *R* to strike the nerve precisely at point *S*, the same humor without being changed will be able to make the rays from point *T* (which is closer) or those from point *X* (which is farther away) come there too. But it will make the ray *TL* go toward *H*, and *TN* toward *G*; and *XL* contrarily, toward *G*, and *XN* toward *H*, and so with the others. Whence in order to represent point *X* distinctly, it is necessary that the whole shape of this humor *LN* be changed and that it become slightly flatter, like that marked *I*; and to represent point *T* it is necessary that it become slightly more arched like that marked *F*.[99] [Cl–44; AT–156]

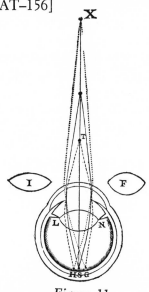

Figure 11

99. E. G. Boring lists five theories that were introduced at various times to explain the eye's adjustments for viewing objects at different distances:

(*a*) The lens shifts forward for nearer, backward for farther, vision (Kepler, 1611; Scheiner, 1619).

(*b*) The lens is made flatter for focusing more distant objects, and vice versa (Kepler identified the principle; Descartes — possibly following a suggestion of Scheiner — asserted that this principle operates in the eye by changes in shape of an elastic lens).

The changes that occur in the size of the pupil serve to moderate the strength of the vision; for it [the pupil] needs to be smaller when the light is too bright, in order that so many rays do not enter the eye that the nerve is damaged thereby; and it [needs] to be larger when the light is too weak, so that enough [rays] enter to be sensed. In addition, assuming that the light remains constant, the pupil must be larger when the viewed object is distant than when it is near; because, for example [Fig. 12], if only just as many rays from point R enter the pupil of eye 7 as are needed in order to be sensed, the same number must enter eye 8, and hence its pupil must be larger.[100] [Cl–45; AT–156]

Smallness of the pupil serves also to render vision more distinct; for you must know that no matter what shape the crystalline humor [lens] may have, it is impossible for it to make the rays that come from different points of the object assemble precisely at correspond-

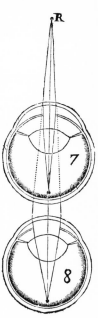

Figure 12

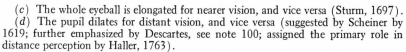

(*c*) The whole eyeball is elongated for nearer vision, and vice versa (Sturm, 1697).

(*d*) The pupil dilates for distant vision, and vice versa (suggested by Scheiner by 1619; further emphasized by Descartes, see note 100; assigned the primary role in distance perception by Haller, 1763).

(*e*) For different distances, the curvature of the cornea changes (Lobe, 1752).

There have been some also who thought that no accommodation for distance perception was needed because the eye has universal focus. See, for the above analysis and further information on this subject, E. G. Boring, *Sense Perception in the History of Experimental Psychology,* New York and London, 1942, 274–278.

As for Descartes, the facts on which he bases his analysis are correct, but today we should phrase the explanation differently, asserting that rays from object point *T* converge to an image point behind the retina (where NG and LH intersect), while rays from point *X* converge to an image point in front of the retina (where NH and LG intersect). For each of these points *T* and *X*, a "blur circle" (Helmholtz) instead of a precise image point is formed on the retina.

100. Modern physiologists would agree with the first of the two points Descartes makes here. Gordon Walls says the size of the pupil must "fix the immediate illumination of the retina . . . at a value above the threshold of stimulation and below the point of dazzlement or injury"; see his *The Vertebrate Eye and its Adaptive Radiation* (Bloomfield Hills, 1942, 153). One could also say that pupil size is so adjusted that it produces an optimal intensity of retinal illumination.

As to Descartes's second point, it is true that in man pupil size is smaller for a nearby object than for the same object farther away, but today we think that this has less to do with the greater luminosity of nearer objects than with the need for increased acuity in the eye accommodated for viewing at close range. Boring (see note 99) correctly notes that the contraction of the pupil for near vision was described by Christopher Scheiner (*Oculus* . . ., 31–41). In the *Dioptrics*, Descartes says that the pupil is small for near objects and for brightly illuminated ones, as well as for objects examined with respect to their details. He says that pupillary adjustment is voluntary — or willed — although we are not ordinarily conscious of the details of the willed response (AT 6:107–108).

ingly different points [at the back of the eye]. Suppose that rays from point R, for example [see Fig. 10], are assembled at point S. Then, of rays from T, only ones that pass [a] through the center and [b] through the circumference of [one of certain] circles that one can describe on the surface of this crystalline humor can assemble exactly at point V. Consequently, the others [other rays from T], which will be as much the fewer as the pupil is smaller, will strike the nerve at other points and cannot fail to cause confusion there. Whence if the vision of one and the same eye is *less strong* at one time than at another, whether from the remoteness of the object or from the weakness of the light, then vision will also be *less distinct*; for, the pupil — being larger when [the light] is less strong — renders vision more confused.[101] [Cl–46; AT–157]

Whence it comes about also that the soul will only be able to see very clearly a single point on the object at a time, namely, that on which all the parts of the eye are trained at that time, other points appearing as much more confused as they are farther from that one. Because, for example, if rays from point R all assemble exactly at point S, those from point X will assemble even less exactly at Y than those from point T assemble at V; and one must judge the same to be true of others in the measure that they are farther from R. But the muscles o,o [Fig. 9] turning the eye very quickly in every direction, serve to remedy this defect: because in no time at all they can apply the eye successively to all points of the ob-

---

101. In this syntactically involved paragraph on optical definition, Descartes makes the following points:

(a) At a given time, the lens can focus on the retina all rays from one point on the optical axis.

(b) Rays from points off the axis will be focused if they pass through an approximately circular line imaginable on the lens surface; so will rays from the object point to the focal point in question.

(c) Other rays from the same object point — especially more peripherally arriving rays — will be sent to various areas of the retina and cause blurring.

(d) When the pupil is constricted, these peripherally arriving rays are excluded and blurring is consequently reduced.

(e) When illumination is weak, the pupil is correspondingly dilated and cannot exclude the peripheral rays; hence, under weak illumination, vision tends to be blurred.

Descartes here appears to foreshadow rather well what certain modern physiologists have termed the sensitivity-acuity seesaw (see Walls, *The Vertebrate Eye* . . ., note 100 above, 69 and 154). Today, however, we tend to think of the blurring of off-axial image points as resulting from the cellular architecture of the retina as much as from peripheral aberration or coma. See W. S. Stiles and B. H. Crawford, "The luminous efficiency of rays entering the eye pupil at different points," *Proceedings of the Royal Society of London*, B112 (1932), 428–540.

ject and thus permit the soul to see all points distinctly one after the other.

I shall not add here in detail what will make it possible for this soul to conceive all differences in color, since I have already spoken of that heretofore [see note 94]. Nor shall I say what objects of vision must be agreeable or disagreeable. For, from what I have explained about the other senses, it is easy for you to understand that light that is too strong must injure the eyes and that moderate light must refresh them. Also that among the colors, green, which consists in the most moderate action (which, by analogy, one can speak of as a ratio of 1 to 2), is like the octave among musical consonances or like bread among the foods that one eats. That is to say, green is the most universally agreeable [color]. Nor finally [will I add here] that all those different fashionable colors which often refresh much more than green are like the chords and passages of a new tune struck up by some excellent lutanist, or like the ragouts of a good cook, which titillate the sense and make it feel more pleasure at first, but which become tedious sooner than do simple and ordinary objects. [Cl–47; AT–158]

It only remains for me to tell you what it is that will give the soul a way of sensing [a] position, [b] shape, [c] distance, [d] size, and [e] other similar qualities, not qualities related to one particular sense (as are those of which I have spoken hitherto), but ones that are common to touch and vision and even in some way to the other senses.[102]

Notice first [Fig. 13], that if hand A touches body C, for example, the parts of brain B from which its nerve filaments come will be otherwise arranged than if it [the hand] touches a body of different shape or size or situated in a different place. It is by this means, then, that the soul will be able to know the situation of the body, and its shape and size and all other like qualities.[103] Similarly, if eye

102. Descartes's first two points on shape, distance, and size will be (a) that the cornea and crystalline humor focus a fair image of the object on the retina (made up of optic nerve endings, he believes), and (b) that different retinal images differently affect the "arrangement" of corresponding parts of the brain. He is not specific here about the nature of the "arrangements" in question, but elsewhere tells us that each sensory nerve filament — or bundle of filaments — can open the entrance to a corresponding pore in the lining of the brain cavity (see notes 103 to 106).

103. Descartes is about to develop an analogy between the hands of a blindfolded person and the eyes of the same person when not wearing a blindfold. The blindfolded person's ability to detect differences in size, shape, and situation depends on

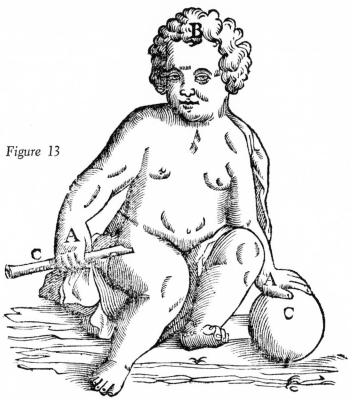

Figure 13

D is turned toward object E [Fig. 14], the soul will be able to know the *position* of this object, inasmuch as [in the brain] the nerves from this eye are differently arranged than if it were turned toward some other object.[104] And [the soul] will be able to know the *shape* [of E], inasmuch as rays from point 1 assembling on the nerve

corresponding differences in the flow pattern of spirits streaming out of the pineal gland (see notes 104 and 130). Descartes makes a similar analysis of the action of the eye in his *Dioptrics*, AT 6:134–141.

104. Descartes makes the point here that the soul (consciousness) derives its impressions directly from physical changes occurring in the brain. What is the nature of these changes? Later, we shall hear that conscious sensory perceptions vary according to the patterns of effluence of spirit particles leaving the pineal gland (see note 130). The speed and direction of outflow depends on two further variables, namely (*a*) the orientation of the gland (which can lean in various directions), and (*b*) the sizes and orientations of the interfibrillar orifices (*entrées*) in the lining of the cavity of the brain. In the *Dioptrics*, it is emphasized that, although a point-to-point image of the object is formed on the retina (and, indeed, in the brain) nevertheless the sort of information the soul receives is merely representative of, not identical to, the actual qualities of the object. Luminosity is translated into the force, and color into the kind, of movements that occur in the optic nerves (AT 6:130–131), and

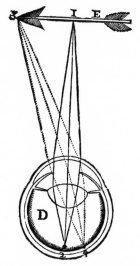

Figure 14

termed optic [the retina] at point 2 — and those from point 3 at point 4, and so forth — will trace there a shape corresponding exactly to the shape of E.[105] Note also that the soul will be able to know the *distance* of point 1, for, as has just been mentioned, in order to make all the rays coming from point 1 assemble precisely at point 2 at the center of the back of the eye, the crystalline humor will be of a different shape than if the object were nearer or farther away.[106] And [note] in addition that the soul will know the distance

these forces and movements are translated into patterns of outflow of spirits from the gland.

105. It is Descartes's idea that variously shaped patterns of retinal activation produce correspondingly shaped activation patterns in the brain (see note 132).

106. *Judgment of distance.* Here Descartes raises issues about the perception of distance. These issues were to be discussed for the next three hundred years; some of them are still being argued today. Here he mentions two cues used in distance perception: the degree of flatness of the lens, and the degree of resolution of the retinal image. As to the former, he suggests that for different degrees of flatness of the lens there must be different degrees of openness of the pores admitting spirits to the nerves that govern the lens's muscles. The differential flow of spirits into these pores permits a judgment concerning the flatness of the lens and hence concerning the distance of the object.

A not dissimilar situation makes possible judgments based on the distinctness of the image. A crucial feature of perception, whether of shape or distance, is comparison; perceptions are based on the drawing of distinctions. In the present case, objects that produce relatively blurred retinal patterns — because they are nearer or farther away than a clearly focused object of fixation — produce, by the same token, relatively blurred patterns of openness in the pores of the ventricular lining and correspondingly

of point 3, and of all others whose rays enter at the same time because, the crystalline humor being properly arranged, the rays from point 3 will not assemble as precisely at point 4 as will those from point 1 at point 2, and so with the others; and their action will be proportionately less strong, as has also been said earlier [see note 96]. And [realize] finally that the soul will be able to know the *size* and all similar qualities of visible objects simply through its knowledge of the *distance* and *position* of all points thereof; just as, vice versa, it will sometimes judge their *distance* from the opinion it holds concerning their *size*. [Cl–50; AT–160]

Notice also [in Fig. 15] that if the two hands *f* and *g* each hold a stick, *i* and *h*, with which they touch object *K*, although the soul is otherwise ignorant of the length of these sticks, nevertheless because it knows the distance between the two points *f* and *g* and the size of angles *fgh* and *gfi*, it will be able to know, as if through a natural geometry, where object *K* is.[107] And quite in the same way [Fig. 16], if the two eyes *L* and *M* are turned toward the object *N*,

confused flow patterns of the animal spirits. Differences in the degree of confusion of the patterns of flow thus serve as cues for judgments of distance. Other cues to distance perception acknowledged by Descartes include: "natural geometry" or triangulation, see note 107; differential luminosity from objects at different distances; and (possibly) pupil size. These cues are discussed more lucidly than here in the *Dioptrics* (AT 6:134–141).

107. *Triangulation in vision.* Certain Stoics are reported by Diogenes Laertius to have said that "the object seen is made known through the intermediation of air extending to it like a stick." That the Stoics said this is confirmed by Galen's objections to their having used the metaphor (*de Plac.*, bk. 7, chaps. 5 [K5, 618–619 and 627] and 7 [K5, 642]).

What Descartes means by "natural geometry" (triangulation) is clearer from the figure than from the text. His idea is not that the same object point is focused on different parts of the right and left retinas but that the whole right and left eyes are differently directed so that their optical axes converge — they cross at the object point. (In very near vision, the individual is slightly cross-eyed.) To permit such "convergence" there must be differences in the flow of spirits from the pineal gland to the right and left sides of the brain ventricle, whence the spirits proceed to nerves that govern the responsible eye muscles. These differences in flow occur automatically, according to Descartes, but the soul is affected by them and obtains information from them about the different patterns of activation of the muscles of the right and left eyes and hence of the eyes' convergence. It does this without recognizing that this is the sort of information it is using.

Much attention has been given in post-Cartesian times to the role of muscularly induced convergence of the eyes in relation to position and distance perception. Do we really make use of kinesthetic information, about the degree of the eyes' convergence? Certainly not for objects at considerable distances. We shall hear from Descartes himself that convergence is ineffective as a cue for objects more than fifteen or twenty feet away, see note 111. In the nineteenth century, Wundt believed in convergence as a cue for spatial localization, but twentieth-century investigations have failed to

Figure 15

the magnitude of line *LM* and of the two angles *LMN* and *MLN* will cause it [the soul] to know where point *N* is.

But it will often enough be possible for the soul to be deceived in all this. For first of all, suppose that the position of hand or eye or brain is forced upon the organ by some external [that is, some other than muscular] cause. [In such a case] the position [of the part] will correspond less exactly to the position of the particles of the brain where the nerves arise than if it depended on muscles alone. Thus the soul, which will only sense [the position or orientation of the part] through the mediation of the parts of the brain, cannot fail to be deceived at such times.[108]

confirm this (see Kenneth N. Ogle, "Perception of distance and of size," *The Eye*, Hugh Davson, ed., vol. 4, New York and London, 1962).

Finally, the sort of muscle sense assumed by Descartes was quite different from the proprioception or kinesthesia known to modern physiology. Whereas for us this information consists in nerve impulses brought to the brain or cord by kinesthetic nerves, for him it consisted in patterns of flow of animal spirits from the pineal gland through the brain ventricles and into the nerves that govern the muscles in question. For theories of binocular vision in the pre-Cartesian era, see A. C. Crombie, "The mechanistic hypothesis and the scientific study of vision . . .," note 90 above, 14 and 22.

108. The muscle physiologist of today would not disagree with the principle Descartes is developing here, that of the deceptive influence of external constraints. An isometrically contracted muscle is more active (its tone is higher) when it is resisting an external force than when no such force is applied, and such differences in isometric activity are relayed to the brain by kinesthetic nerves. (The possibility of "deception"

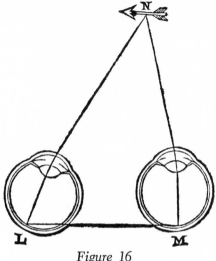

Figure 16

Suppose for example [Fig. 17] that hand *f*, being disposed in itself to be turned toward *o*, finds itself constrained by some external force to remain turned toward *K*. In that case, the parts of the brain whence the nerves come will not be arranged in quite the same way as they would be if it were by muscular force alone that the hand was thus turned toward *K*. Nor will they be arranged as they would

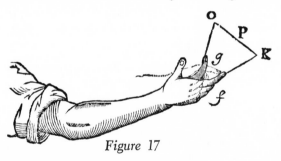

Figure 17

is suggested in a simple experiment in which a standing person pushes against a door jamb with the back of one hand, say the left, while both arms are held at the sides. When he stops pushing, his left arm reflexly swings outward for a moment.)

Descartes make a considerable point, in this and the following paragraphs, of the deception involved in what we term isometric contractions against an external resistance. His idea is to show that the soul gets its information about the direction of an object from the *muscles* that orient the eyes (or hands) toward that object, or more precisely, from the nerves that effect that orientation. In a soulless animal this information could elicit the appropriate response even in the absence of awareness. See also, on muscular constraint, the *Dioptrics*, AT 6:141.

be if the hand were in fact turned toward o. They are rather arranged in a manner intermediate between the two, that is, as if turned toward P. And thus the arrangement that this constraint will give to the particles of the brain will cause the soul to judge that the object K is at point P and that it is a different object from that which is touched by hand g. [Cl–51; AT–161]

Similarly [in Fig. 18], if eye M is turned away by force from object N, and arranged as if looking toward q, the soul will judge that the eye is turned toward R. In this situation rays from object N will enter the eye in the way that those from point S would do if the eye were in fact turned toward R; hence it [the soul] will believe that this object N is at point S and that it is a different object from the one being looked at by the other eye.[109]

Similarly [in Fig. 19], the two fingers t and v, touching the little ball X, will make the soul judge that they are touching two different things because they are crossed and kept forcibly in an unnatural position. [Cl–52; AT–161]

Moreover, if the rays (or other lines through whose mediation the actions of distant objects pass toward the senses) are curved, the soul, which would ordinarily suppose them to be straight, will

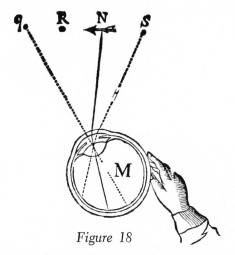

Figure 18

109. Some of the effects described here occur when we rub one eye or push it gently. To test Descartes's contention about the eye under muscular constraint, write the letter N on a blank paper, and then the letter S about an inch below and slightly to the left. By gentle pressure on the outer edge of the right upper eyelid one can dissociate the single image of N into two images, one of which (that from the right eye) can be superimposed on the (left eye's) image of S.

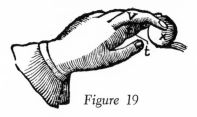

Figure 19

have occasion to be deceived. As, for example [Fig. 20], if stick *HY* is curved toward *K*, it will seem to the soul that object *K* which this stick touches is in the direction of *Y*. And if eye *L* [Fig. 21] receives rays from object *N* through a glass *Z* [a prism] that curves them, it will seem to the soul that this object is in the direction of *A*. Again, suppose [in Fig. 22] that eye *B* receives rays from point *D* through glass *C* [a magnifying glass], which I assume bends them as though they were coming from point *E* and bends those from point *F* as though they were coming from point *G*, and so with other rays. [In such a case] it would seem to the soul that the object *DFH* is as far away and as large as *EGI* would seem to be.[110] [Cl–53; AT–162]

To conclude, it must be noted that none of the soul's means of knowing the distance of objects will be quite sure, for the following

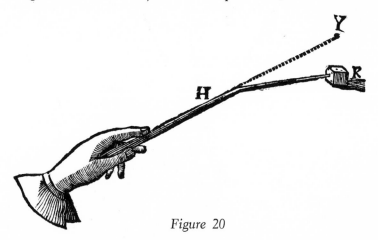

Figure 20

110. With eyes closed, the subject will locate an object wrongly if he thinks the curved *stick* he is using is straight. With eyes open, he will locate it wrongly if he thinks the curved *light rays* are straight. In both cases he is operating in terms of the natural geometry mentioned above in note 107. The refractive illusions referred to can be used remedially to correct defective vision.

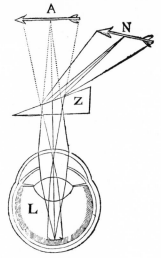

Figure 21

[three] reasons. [First,] as for angles like *LMN* and *MLN* and so forth [see Fig. 16], these no longer change appreciably with distance for objects fifteen or twenty feet away or more. [Second,] as for the shape of the crystalline humor, it changes even less appreciably than the foregoing for objects more than two or three feet from the

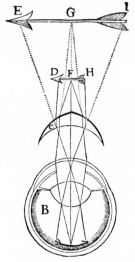

Figure 22

eye.[111] And [third and] finally, what about judging the distances of objects from one's opinion of their size, or from the fact that rays from different points of the object differ in the precision with which they assemble at the back of the eye? The example of *tableaux de perspective*[112] shows us amply how easy it is to be deceived. For if [peculiarities of] shape make us overestimate the size of visual objects, or if their colors are somewhat obscure, or their outlines somewhat indefinite, these things make them appear to be more distant, and larger, than they actually are. [Cl–55; AT–163]

Now having explained to you the five external senses as represented in this machine, I must tell you something of certain internal senses it contains.[113]

When the liquids which serve, as mentioned earlier, as a sort of aqua fortis in the stomach, and which enter ceaselessly from the

111. Limits are specified here for the effectiveness of convergence (training of the eyes on nearby objects) and accommodation (lens shape) as cues in distance perception.

112. This phrase was sometimes applied to various devices employed in painting and drawing to permit two-dimensional representations of three-dimensional objects, sometimes to paintings in which perspective principles were employed (Descartes had already used the phrase in the *Dioptrics*, AT 6:147). During the century preceding Descartes, a number of books on perspective appeared, some emphasizing its connection with optics, others its applications to art. Thus the year 1535 saw publication of the *Optics . . . Commonly Called Perspective* of the thirteenth-century Polish mathematician Witelo (*Peri optikes . . . quam vulgo perspectivam vocant*, Nuremberg); and the year 1572 saw the Latin translation of the *Thesaurus of Optics* of the tenth-century Arab physician Alhazen (Hasan ibn Hasan) (*Optici Thesaurus . . .*, Basel) (see note 90 above). Descartes was aware of these works, but it is difficult to know how familiar he was with authors who treated perspective more directly in relation to graphic representation — for example, Jan de Vries (1560), J. Cousin (1560), Danielo Barbaro (1569), M. Bassi (1572), Salomon de Caus (1612), S. Marolais (1614), and H. Houdin (1625).

113. *External vs internal senses*. This distinction had been conventional since medieval times, but Descartes radically altered the accepted list of internal senses. The external ones had been, for almost all writers on the subject since Greek antiquity, the classical five. But in medieval philosophy and Renaissance physiological theory, the internal senses were usually given as: the common sense plus, depending on the author, some combination of discrimination, imagination or fantasy, memory, and even reason (see notes 125 and 137). Hunger and thirst, internal senses to Descartes, were not included.

In his *Principles*, Descartes specifies two internal senses: first, a sense comprising "hunger, thirst, and all other natural appetites . . . excited in the soul by the movements of the nerves of the stomach, throat, and all other parts that serve natural functions and for which such appetites exist"; and second, a sense comprising "joy, sadness, love, anger, and all the other passions which depend chiefly on a little nerve that goes to the heart but also on nerves that go to the diaphragm and other internal parts of the body" AT 8:316 (9:311). See also note 53 above and, for treatments less systematic than the present one, *Dioptrics* (AT 6:109), *Meditations* (AT 9:25), and letter to Elizabeth (AT 4:310).

whole mass of the blood through the extremities of the arteries, do not find there enough food to dissolve so as to employ their whole force, they turn against the stomach itself and, agitating the filaments of its nerves more strongly than is usual, they cause motion in the parts of the brain from which the filaments come. This will cause the soul, when united to this machine, to conceive the general idea of *hunger*. And if these liquids are so disposed as to employ their action against certain particular foods rather than others, much as ordinary aqua fortis dissolves metals more easily than wax, they will act in a particular fashion also against the nerves of the stomach. This will cause the soul at such times to conceive an appetite to eat certain foods rather than others. (*Hic notari potest mira huius machinae conformatio, quod fames oriatur ex jejunio: sanguis enim circulatione acrior sit, & ita liquor ex eo in stomachum veniens nervos magis vellicat, idque modo peculiari, si peculiaris sit constitutio sanguinis: unde pica mulierum.*) [114] Now these liquids are assembled chiefly at the bottom of the stomach, and it is there that they cause the feeling of hunger. [115] [Cl–56; AT–164]

But many part[icle]s [of the gastric fluid] rise continuously toward the gullet, and when they do not come there in sufficient numbers to moisten it and fill its pores in the form of water [saliva

114. "That hunger arises from fasting is witness to the wonderful conformation of this machine: for the blood in circulation [at such times] would be more acrid and the juice coming thence to the stomach would thus irritate the nerves to a greater extent — and in a peculiar manner if the constitution of the blood were peculiar; from the same cause [comes also] pica [the inordinate pangs] of [pregnant] women." La Forge considered this Latin sentence extraneous to the French text of Descartes; he thought it was a marginal comment added by some reader and later erroneously inserted into the text by "an unintelligent or overly faithful copyist," AT 11:164.

115. Galen had related hunger to the *repletion* that the body must experience to offset its continuous *depletion*. Depletion occurs, for example, through perspiratory loss from organs just under the skin. Such depletion arouses the unconscious natural (as opposed to conscious animal) appetite of the parts. The parts draw repletion from the veins, and the veins — by suction — from the stomach. Hunger is the stomach's "animal" (conscious, nerve-mediated) sense of its desire to compensate (by taking food) for the suction exerted on it by the veins (*Sympt.*, bk. 1, chap. 7, K7, 130–131). Galen described the gastric nerve plexuses and said they extend the sense of hunger to the stomach (*UP*, bk. 4, chap. 7, K3, 277, and bk. 9, chap. 11, K3, 727).

An examination of Vesalius (*Fabrica*, 493), Columbus (*Anatomica*, 227), Piccolhomini (*Praelectiones*, 403–414), Bauhin (*Theatrum*, 314), and Riolan (*Oeuvres*, 286) suggests that early Renaissance theory on hunger contained few innovations. The most extended analysis among the foregoing was that of Piccolhomini, but it was an elaboration on Galenic assumptions and apparently had no special influence on Descartes if he knew of it. Descartes's theory has a rather modern ring, because of

— see note 36], they rise instead in the form of air or smoke. At such times, acting against its nerves in an unusual fashion, they cause a movement in the brain that will make the soul conceive the idea of *thirst*. [Cl–57; AT–164]

Similarly, when the blood that goes into the heart is more pure and subtle and is kindled more easily than usual, this arranges the little nerve that is there in the manner required to cause the sensation of *joy*. And when this blood has quite contrary qualities, [it arranges the nerve] in the manner required to cause the sensation of *sadness*.[116]

its prescience with respect to a particle-based idea of digestion. But, contrary to his view, hunger has not turned out to be caused by unemployed gastric juice. He speaks briefly of hunger and thirst also in the *Principles*, AT 8:316 (9:311).

116. *Joy and sadness.* Medical theorists had long sought the physiological basis of these states. Correlations of (*a*) the blood condition with (*b*) emotional states (and emotional dispositions, see note 117) were frequent, but the nature of the causal relations between the two had been disputed.

In antiquity, the Hippocratic treatise *The Sacred Disease* had made two humors, phlegm and yellow bile, responsible for joy and sadness but insisted that they act on the brain as seat of intelligence and emotion and not on the heart — although the heart is involved, we are told, in body changes that accompany emotions (*Oeuvres complètes* . . ., E. Littré, ed., Paris, J. Baillière, 1839 to 1861, 6:387–395). Galen associated sadness with heart pains due to an excess of black bile (*de Plac.*, bk. 2, chap. 8, K5, 273). He noted that whereas Praxagoras (circa 300 B.C.) and Aristotle had made the heart the site of origin of nerves in general, the heart has only one small nerve pair connecting it with the brain via the two "sixth" (our tenth or vagus) cranial nerves. Galen probably thought the heart nerves extend the sensory power of the brain to the heart, permitting conscious cardialgia (*de Plac.*, bk. 1, chap. 10, K5, 205–210; *Anat. Adm.*, bk. 7, chap. 8, K2, 613; and *UP*, bk. 16, chap. 5, K2, 289). Vesalius, too, acknowledged small nerve connections to the heart from the "sixth" pair and asserted that these were "for the sake of sensing distress" (*tristantium* . . . *sensus gratia*) and not for moving the heart (*Fabrica*, 588 and 596; see also note 47 above and 122 below). Paré thought that the sensation of a pleasurable object could influence the heart to become enlarged and send more blood, spirits, and heat to various organs including the face, which therefore manifests pleasure; unpleasant objects have inhibitory effects on the heart; joy and sadness are the movements through which the distribution of blood, spirits, and heat are altered in the manner described (*Oeuvres*, 36). See also, for Renaissance treatments of the nerves of the heart, Rothschuh, *D.*, 95, note 2.

Descartes's model parallels Galen's at least to the extent that the nerves of the heart are involved. In the *Principles* (AT 8:316–318 [9:311–312]), Descartes says that very pure and well-tempered blood, by causing the heat to beat more easily and forcefully than usual, moves the heart's nerves in such a way as to instigate in the brain (and in the soul) a joyful sensation. Or some benefit imagined by the brain may stimulate the heart — through motor nervous action — to beat easily and forcefully, and this beating will reciprocally stimulate the brain (and the soul) — through sensory nervous action — to experience feelings of joy. Awareness of sadness arises from contrary conditions of the heart.

In the *Passions*, we hear that "the consideration of a present good excites joy in us and the consideration of a present evil, sadness — if the good or evil be repre-

From this you can well enough understand what there is in this machine that corresponds to all the other internal sensations in us; whence it is time that I commence to explain to you [a] how the animal spirits pursue their course in the cavities and pores of its brain, and [b] what funtions depend upon them.

If you have ever had the curiosity to look closely at the organs in our churches, you know how their bellows push air into certain receptacles called — for this reason, presumably — wind trunks. [You know] also how from there the air enters the pipes, now one, now another, as the organist moves his fingers on the keyboard. And you can think of the heart and arteries of our machine (which push animal spirits into the cavities of its brain) as similar to the bellows (which push air into the wind trunks of organs); and of external objects (which, by displacing certain nerves, make spirits from the brain cavities enter certain pores) as similar to the organist's fingers (which, by pressing certain keys, make air from the wind trunks enter certain pipes).

Now the harmony of the organ depends not at all on the externally visible arrangement of the pipes nor on the shape of the wind trunks or other parts, but only on three things, namely, [a] the air

---

sented as pertaining to us" (AT 11:376). (This sentence is an almost verbatim quotation from Galen, de Plac., bk. 4, chap. 2, K5, 366.) In the Passions as in the Principles, the experience of joy or sadness depends upon sensory stimuli transmitted to the brain by the nerves of the heart (AT 11:405–406). In the Passions, Descartes distinguishes corporeally instigated joy from pure intellectual joy "which occurs in the soul purely through the action of the soul," although he stresses that intellectual is almost always accompanied by corporeal joy (AT 11:396–397).

He also elaborates in the Passions an idea mentioned in Man (note 67 above), namely that stimuli that would injure unsoundly constituted nerves can be pleasant just because they testify to the soul that the constitution of the nerves is sound; the realization of this soundness is joy. He draws an analogy between this situation and the one that prevails in a theater where we can take pleasure in emotions that elsewhere would prove unpleasant (Passions, AT 11:376 and 396–401).

Finally, in the Passions, Descartes seems to view joy and sadness as responses to sensory nerve action throughout the body; he no longer limits them to activities of the nerves of the heart (a limitation about which he had been quite firm in an earlier unpublished note, Anatomical Excerpts, AT 11:595).

In a remarkable speculation contained in a letter written late in life to Chanute, Descartes suggests that the first passion of the soul when joined to the machine would be joy (because the soul would first of all sense the soundness of the body); but (since the body must replace what it continuously displaces) the soul would soon acquire a second passion, namely a desire — and love — for food; next would follow sadness (if the food was not forthcoming); and finally hate (if, for example, what was forthcoming was bad) (AT 4:604–605). These four passions now seem to Descartes to be the only innate ones.

that comes from the bellows, [b] the pipes that sound, and [c] the distribution of this air to those pipes. And let me call to your attention that, here too, the functions under consideration in no wise depend on the external shape of the visible parts which the anatomists distinguish in the substance of the brain nor on the shape of its cavities, but only [a] on the spirits that come from the heart, [b] on the pores of the brain through which they pass, and [c] on the way in which these spirits are distributed to these pores. Whence it is only necessary that I explain to you in proper order what is of most importance in connection with these three things. [Cl–58; AT–166]

Firstly, as to animal spirits, they can be more or less *abundant*, and their part[icle]s can at different times be more or less *coarse*, more or less *agitated*, and more or less *uniform* [in size, shape, and force — see below]; and it is by means of these four differences that all of the various humors or natural inclinations present in us are also represented in this machine (at least insofar as these do not depend on the constitution of the brain or on particular affections of the soul).[117] For if these spirits are unusually abundant, they are appropriate for exciting movements in this machine like movements that give evidence in us of *generosity*, *liberality*, and *love*. And [they

117. The theory offered here by Descartes is his contribution to a long and complex history of ideas relating psychological and emotional differences ("natural inclinations") to differences in constitution. In antiquity these ideas began with the pre-Socratics. Empedocles believed that thought and mind reside in blood and differ among individuals according to the size, spacing, and mixture of their blood particles (see F. Solmsen, "Tissues and the soul," *Philosophical Review*, 59 [1950], 438). In the Hippocratic treatise *On Regimen*, various blendings of constitutive fire and water serve to place men on a scale with superior rationality at one end and madness at the other (bk. 1, chap. 35). Galen wrote a short but solid treatise entitled *That Habits of Mind Depend on the Temperament* [*krasis*, blend] *of the Body*. Such early theories initiated a flood of later speculations — too copious to be epitomized here — on the constitutional basis of psychological and/or emotional types.

Descartes's idea — that these variations depend on differences in the dynamics of the particles of the animal spirit — illustrates the dangers inherent in his too deductive, too systematic explanatory procedures. Yet his conclusions seem, in retrospect, as legitimate as those inherent in the humoral psychotypology of medieval and Renaissance psychology (the concept that people tend to be sanguineous, phlegmatic, choleric, or melancholic). Gilson (*ISC*, 103) offers, as a possible stimulus to Descartes's association of passions with spirits, a commentary by the Coimbran fathers on Aristotle's *On Youth and Old Age*; but Aristotle himself did not mention *pneumata* (spirits) in that treatise. The idea that animal spirits differ in regard to particle size is further developed by Descartes in the *Passions* (AT 11:339–340). See also Rothschuh, D., 97, notes 1 and 2.

excite movements that give evidence] of *confidence* or *courage* if their part[icle]s are unusually strong and coarse; and of *constancy* if, in addition, they are unusually uniform in shape, force, and size; and of *promptness, diligence,* and *desire* if unusually agitated; and of *tranquility of spirit* if unusually uniform in their agitation. Whereas, on the contrary, if the same qualities are lacking, these same spirits are appropriate for exciting movements in [the machine] entirely like movements in us that bear witness to *malice, timidity, inconstancy, tardiness,* and *ruthlessness.* [Cl–59; AT–167]

And know that all the other humors or natural inclinations are dependent on those mentioned above. Thus the *joyous humor* is composed of promptitude and tranquility of spirit, and generosity and confidence serve to make the joyous humor more perfect. The *sad humor* is composed of tardiness and restlessness and can be augmented by malice and timidity. The *choleric humor* is composed of promptitude and restlessness, and malice and confidence fortify it. Finally, as I have just said, liberality, generosity, and love depend upon an abundance of spirits, and form in us that humor which renders us complaisant and benevolent to everyone. Curiosity and the other impulses depend upon the agitation of the part[icle]s of [the animal spirits]; and so with the other inclinations.[118]

But because these same humors or at least the passions to which they predispose us are also very dependent on the impressions that are made in the substance of the brain, you will be able to understand them better hereafter; and I shall content myself here with telling you the causes whence differences in spirits arise. [Cl–60; AT–167]

The juice of the food that passes from the stomach into the veins[119] on being mixed with the blood always communicates some

118. It is difficult to be sure to what extent Descartes intends any parallel between the three humors mentioned here (joyous, sad, choleric) and the four temperaments of medieval medical and psychological thought (sanguineous, phlegmatic, choleric, melancholic). His scheme, in any case, is entirely different in conception, being corpuscular in its assumptions rather than humoral; see note 117.

119. On what grounds does Descartes postulate absorption from the stomach? Galen saw partially concocted food as drawn to the liver from both the stomach and the intestine (*UP*, bk. 4, chaps. 2, K3, 269 and 17, K3, 323). He had a reasonably acceptable idea of the tributaries of the portal vein in animals (see for example, *De venarum arteriarumque dissectione*, K2, 780–785). Vesalius, focusing on man, gave a much more precise and elaborate picture (*Fabrica*, 262–267). His reconfirmation of

of its own qualities thereto and, among other things, usually makes it more coarse when it first mixes freshly therewith. Whence, at this time, the particles of blood that the heart sends to the brain to constitute the animal spirits are generally not so agitated, strong, or abundant [as they are at other times]. Consequently they do not usually make this machine so nimble or quick as it becomes a while after digestion is finished and after the same blood, having passed and repassed through the heart several times, has become more subtle [see note 22].

The air of respiration, likewise, being mixed in some way with the blood before it enters the left cavity of the heart, makes the blood kindle more strongly,[120] and produces more lively and agitated spirits [in the heart] in dry weather than in humid weather: just as flames of every sort are found at such times to be more ardent.

When the liver is well disposed and elaborates perfectly the blood that goes to the heart, the spirits that leave this blood are correspondingly *more abundant* and *more uniformly agitated*. And

---

gastric tributaries to the portal vein added weight to the belief that absorption from the stomach occurs. He and many of his successors — including Columbus (*Anatomica*, 227–229), Piccolhomini (*Praelectiones*, 108), du Laurens (*Oeuvres*, 188), and Riolan (*Oeuvres*, 272) — agreed that food (chyle) proceeded to the liver from both the stomach and intestine. There was some discussion of the route, however, related partly to doubts raised by Galen's belief (*UP*, bk. 4, chap. 19, K3, 336) that the same veins could carry both chyle to the liver from the stomach and intestines, and concocted blood in the reverse direction. (Until Harvey's time, the veins were supposed to distribute nutriment from the liver to the whole body directly without its necessarily passing first through the heart.) Today, the stomach is considered to play only an insignificant role in the absorption of nutrients.

120. *Blood and air*. Galen thought that some of the blood that comes to the lungs passes over (by way of arteriovenous anastomoses) to the pulmonary veins. He also believed that the pulmonary veins carry air, some of it already converted in the lungs to animal spirits, the rest destined to undergo this conversion upon arrival in the heart (*UP*, bk. 7, chap. 8, K3, 541). See also R. E. Siegel, *Sudhoff's Archiv*, (1962), 311–322. The idea that the pulmonary veins contain both air and blood was therefore Galenic, and by the same token was part of the neo-Galenic tradition with which Descartes was thoroughly conversant.

In Harvey's scheme, virtually *all* of the blood arriving in the lungs is returned to the heart by the pulmonary veins (some parts of it being withdrawn to nourish the lungs). The blood passes through pores in the lungs, he said, "so that inhaled air may temper it and protect it against boiling and suffocation" (*Motion of the Heart*, end of chap. 6, retranslated from the Latin). Harvey adopted the idea of the pulmonary circulation chiefly from Columbus, who had argued that the pulmonary arteries were unnecessarily large to be thought of as merely nourishing the lungs; thus the blood they contain must pass through the lungs, become mixed with air, and thence goes back to the heart for distribution to the body in general (*Anatomica*, 177–178 and 222–224). Descartes's hypothesis is eclectic. For him the lungs exert a cooling and condensing effect, but they also add a certain amount of air to the blood.

should the liver happen to be incited by its nerves, the subtlest part of the blood it contains, rising directly to the heart, will produce spirits correspondingly *more abundant* and lively than is usual — though *not so uniformly agitated.*

If the gall [bladder], which is intended to purge the blood of those of its parts that are *most suited* to be enkindled in the heart, fails in its task, or if being contracted through [the action of] its nerve it regorges into the veins the matter it contains, then the spirits will be, to that extent, *more lively* and *more unevenly agitated* withal.

Per contra, if the spleen, which is intended to purge the blood of parts *least suited* to be enkindled in the heart, is ill disposed, or if, under pressure from its nerves or from any other body whatever, it regorges into the veins the matter that it contains, then the spirits will be to that extent *less abundant,* and *less agitated,* and *less uniformly agitated withal.*[121] [Cl–61; AT–169]

In sum, whatever can cause any change in the blood can also cause change in the spirits. But above all, the little nerve that ends in the heart is able to dilate and contract both [*a*] the two entrances

121. *The gall bladder and spleen.* The abundance and agitation of spirits depends on the ratio of flammable to nonflammable matter brought to the heart in the blood. Normally the gall bladder removes (part of) the flammable component of the blood, whereas the spleen removes (part of) the nonflammable component. But if these organs are malfunctioning (or are unusually affected by their nerves), the opposite of their usual influence prevails.

Galen believed that as chyle enters the liver it is resolved into three components respectively comparable to wine (blood), the foam or flower of the wine (yellow bile), and the leas (black bile) (see note 19). Yellow bile is attracted to the gall bladder, altered, partially assimilated, the rest sent to the intestine. Black bile is attracted and assimilated by the spleen, some of it going on to the stomach to promote the retention of the food (*UP*, bk. 5, chap. 4, K3, 358).

A number of points were debated by pre-Cartesian Renaissance medical writers: Is the bile attracted by the gall bladder, or repelled by the liver, or both? What is the disposition of the ducts? Is there a separate duct connecting the gall bladder with the stomach? Does the gall bladder utilize bile as food for itself — or merely control its flow to the duodenum? Almost all of the major authors of the period — Fernel, Columbus, Falloppius, Bauhin, du Laurens, Riolan — entered the debate, but its details lie outside our present concerns, since Descartes was not much influenced by them.

With respect to the spleen, the pre-Cartesian anatomists were more explicitly anti-Galenic. There was no unanimity, but there was a tendency to assign the spleen a special role in sanguification (see especially Bartholin, *Institutiones*, 79, and his account of the views of Rondelet). Descartes was not improbably aware of these ideas, but his scheme is very much his own. Rothschuh (*D.,* 97–99, notes) sees Descartes as reflecting in these passages the influence of the humoral pychotypology, then popular, according to which people were psychologically as well as physically sanguineous, phlegmatic, choleric, melancholic; on which see also note 118 above.

through which the blood of the veins and air of the lung descend, and [b] the two exits through which blood is exhaled and driven into the arteries. [Hence this nerve] can cause a thousand differences in the nature of the spirits: just as the heat of certain enclosed lamps which the alchemists use can be moderated in several ways according as one opens, to a greater or less degree, now the conduit through which the oil or other aliment of the flame comes in and now that by which the smoke goes out.[122]

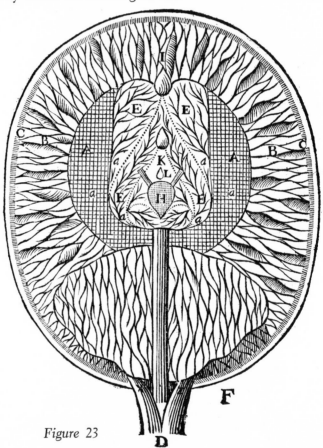

Figure 23

122. In other words, the properties of the blood — and those of particles of blood that will later be separated as spirits — depend upon the mixture of venous blood with air. On the innervation of the heart, see notes 53 and 116. We have already heard Galen compare the heart to a lamp (note 24). Earlier, Aristotle had drawn a similar comparison between the heat of the heart and a fire whose coals flame differently depending on the degree to which they are damped (*Juv.*, 469b7–20).

Secondly, concerning the pores of the brain, they must be imagined as no different from the spaces that occur between the threads of some tissue [for example, a woven or felted fabric];[123] because, in effect, the whole brain is nothing but a tissue constituted in a particular way, as I shall try to explain to you here. [Cl–62; AT–170]

[In Figs. 23 and 24] conceive surface AA facing cavities EE to be a rather dense and compact net or mesh, all of whose links are so

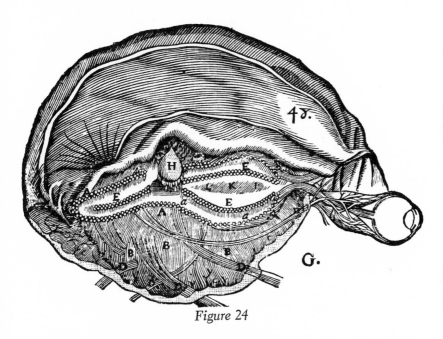

Figure 24

123. *"Tissue."* Descartes does not use *tissu* in the modern sense, but rather to suggest the fibrous mesh or weave of the parts. Thus, Schuyl in his Latin translation of *Man* renders the term by *textum* and *textura* (something woven).

While not strictly relevant to our interest in Descartes, the subsequent history of these terms is of interest. After his time, a more usual term than *textum* for the body's supposed fibrous mesh was *tela*. For example, von Haller used the expression *tela cellulosa* (primarily in connection with what we see as areolar connective tissue — although he thought of a comparable mesh as forming a common basis for all the solid parts of the body), and von Haller's French translator rendered *tela* as *tissu* (for example in *Primae lineae physiologiae . . .*, Göttingen, Vandenhoeck, 1747, 4, 5, and 7–11; trans., *Élémens de physiologie . . .*, Paris, Guillyn, 1761, 2, 3 and 4–6). The term *tissu* was subsequently used by Haller's critic Bordeu (for instance, in his *Recherches sur le tissu muqueux, ou l'organe cellulaire . . .*, Paris, 1767), and, partly under Bordeu's influence, by Bichat who around A.D. 1800 laid the basis for modern tissue theory.

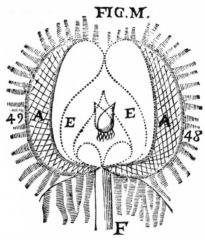

Figure 25

many little conduits which the spirits can enter and which, always facing toward gland *H* [the pineal] whence these spirits emanate, can easily turn hither and thither toward different points on this gland — as you see that they are turned differently in the 48th than in the 49th diagram [right and left sides of Fig. 25]. And assume that from each part of this net arise several very thin threads of which some as a rule are longer than others; and that after these threads have been differently interlaced through the space marked *B*, the longer of them descend toward *D*, and from there, comprising the marrow of the nerves, proceed to spread through all the members [see Fig. 24].[124] [Cl–64; AT–171]

124. From this and other paragraphs, we learn that the brain substance consists of intermeshed fibers extending outward from the lining of the brain ventricles, in some cases to the arterioles of the pia mater and in some cases beyond the brain to the various organs of the body. Bundles of such filaments extending beyond the brain (as cranial nerves and spinal cord) are enclosed within sleeve-like derivatives of the membranes of the brain (see note 46).

Descartes regards the interfilamentous spaces in the brain as channels that conduct the animal spirits from the ventricles to the nerve tubules with which they are continuous. The essence of motor control is, then, the direction of animal spirits into the proper interfilamentous channels for transmission to the proper nerve. Next we shall hear the manner in which this directing of spirits occurs in the case of willed as well as of involuntary or reflexive responses. We shall also be given further details of the subvisible structure of the brain, but on the whole Descartes is cryptic on this subject. Some help in understanding his concept is obtainable from La Forge's commentary (*l' Homme*, 295–325).

Assume also that the chief characteristics of these filaments are [a] that they can be flexed rather easily in all sorts of ways merely by the force of the spirits that strike them, and [b] that they can retain, as if made of lead or wax, the flexure last received until something exerts a contrary pressure upon them.[125]

Finally, assume that the pores we are considering are nothing but the intervals between these threads and [that they] can be diversely enlarged and constricted by the force of the spirits that enter them according as that force is more or less strong and [according as the spirits] are more or less abundant; and that the shortest of these threads betake themselves to the space cc [Fig. 23],[126] where each terminates against the extremity of one of the little vessels that are there and receives nourishment from it. [Cl-65; AT-171]

Thirdly — in order to explain all the particularities of this tissue more conveniently, I must begin to speak to you now about the distribution of these spirits.

The spirits never stop for a single moment in any one place, but as fast as they enter the brain cavities EE [Figs. 23 and 24], through apertures in the little gland marked H, they tend first toward those tub[ule]s, a and a, which are most directly opposite them; and if those tubules are not sufficiently open to receive them all, they receive at least the strongest and liveliest of the particles thereof, while the feeblest and most superfluent particles are pushed aside toward the conduits I, K, L which face the nostrils and the palate. Specifically, the most agitated [are pushed] toward I, through which — if they still have much force and do not find the passage free enough — they sometimes pass out with so much violence that they tickle the internal parts of the nose which causes *sneezing.* Then other [particles are pushed] toward K and L, through which they can leave quite easily because the passages there are very large;

125. La Forge (*l'Homme*, 297) compares the fibers ending at the internal brain surface to the straws in a whisk broom; but the brain fibers are flexible and align themselves with their free ends roughly facing the pineal gland under the influence of currents emanating therefrom. Further on, we shall hear Descartes make the flexures or folds the nerves receive the basis of imprinting (to which he refers variously by the French verbs *tracer* and *imprimer,* and the noun *impression*) and memory (see note 137).

126. Probably the subarachnoid space of the brain envelope, or its cisternae, the "little vessels" being those with which the pia mater is well supplied.

or if they fail to do so, being forced to turn back toward tubules *a* and *a* in the internal surface of the brain, they promptly cause a *dizziness* or *vertigo* which disturbs the functioning of the *imagination*.[127] [Cl–66; AT–172]

And note in passing that the weaker part[icle]s of the spirits come less from the arteries inserted in gland *H* [the pineal] than from those which divide into a myriad of very small branches and thus carpet the cavities of the brain.[128] Note also that these particles can

---

127. From one point of view, these "explanations" seem gratuitous — the fatal consequence of Descartes's rationalistic, deductive procedures. We have to understand them, however, in terms of his larger purpose, which was to specify the subvisible mechanics of a hypothetical robot equipped to do everything that the human body can do. More important than what Descartes says here about sneezing and dizziness is his general point that spirits tend to flow copiously from the blood into the pineal gland, and from the gland into the ventricular cavities of the brain, and thence into the pores of interstices of the brain substance. The flow of spirits through the ventricles, and the alterations in this flow, are basic features of Descartes's theory of the brain and of his views about the nature of ideas. Ideas differ, he claims, according to two specific conditions, namely the position of the pineal gland and the differential effluence of spirits from different parts of its surface (see note 130).

128. *Origin and number of spirits.* On this subject we noted earlier that Galen viewed animal spirits as concocted out of vital spirits in the blood and as discharged into the brain cavities by "exhalation" (*anathymiasis*). He believed that the discharge occurred from the blood in both the rete mirabile (which lies outside of and below the brain) and the choroid plexus (which lies within the ventricles). For a discussion of this subject, see Siegel, *Galen's System*, 109–115, and the references of these two structures in the index of May's translation of *UP*.

Vesalius notoriously denied the existence of a rete mirabile in man and charged Galen with improper application to man of structures found only in animals (*Fabrica*, 310, 524, and 624). During the post-Vesalian period, anatomists whose work was known to Descartes in general accepted the choroid plexus as a source of animal spirits, in most cases without denying the existence of the rete. The rete was acknowledged — and associated in various ways with the choroid plexus — by Fernel (*Physiologie*, 381–383), Paré (*Oeuvres*, 1579, 172), Piccolhomini (*Praelectiones*, 253), Bartholin (*Institutiones*, 220), Riolan (*Oeuvres*, 588–590), and many others.

As to the number of vital spirits — whether one, two, or three — Galen was sure of two (animal and vital) and tentative about the third (natural) (see O. Temkin, "On Galen's pneumatology," *Gesnerus*, 1951, 180–189, and May, *Galen*, 48–49). Most Renaissance authors accepted three spirits (Fernel, *Physiologie*, 381–383; Paré, *Oeuvres*, 1579, 21–22; Piccolhomini, *Praelectiones*, 12; Bartholin, *Institutiones*, 70, 118, and 146; du Laurens, *Historia*, 314). But Vesalius seems not to have mentioned the natural spirits; and Columbus explicitly rejected them (*Anatomica*, 164, 175, and 191). Argenterio acknowledged only one spirit (*De somno et vigilia . . .*, Florence, 1556, 305–310), and du Laurens refuted this opinion (*Historia*, 544–546). Although Harvey more or less dispensed with spirits and made them indistinguishable from other parts of the blood, he presented his analysis in later works that could not have influenced *Man* (see his *Circulation of the Blood*, 37–41, and W. Pagel, *William Harvey's Biological Ideas*, Basel and New York, 1967, 252–255).

This summary suggests the instability of pneumatic theory at the time of Des-

easily thicken into phlegm. Only in grave illness, however, do they do this in the brain itself; ordinarily it occurs in those large spaces beneath the base of the brain between the nostrils and the gullet,[129] just as smoke converts easily into soot in the flues of the chimney but never in the hearth where the fire burns.

Note also that when I say that spirits leaving the gland tend toward the most directly "opposite" regions of the internal surface of the brain, I mean merely that they tend where the arrangement of the brain at the time impels them, not necessarily to regions that face them rectilineally.[130] [Cl–67; AT–173]

Now the substance of the brain being soft and pliant, if no spirits entered its cavities these cavities would be very narrow and almost entirely closed, as they appear in the brain of a dead man. But the source that produces these spirits is ordinarily so copious that they have a capability, corresponding to the amount of them entering the cavities, to push outward in all directions the matter that surrounds them, thus causing this matter to expand and tighten all nerve filaments that arise there [Fig. 26]: just as the wind when somewhat strong can inflate the sails of a ship and tighten all the ropes to which the sails are attached. Whence it follows that at such

---

cartes. It shows the variations on Galenic doctrine with which he was undoubtedly familiar and among which he was free to move in developing a characteristically corpuscular and mechanical pneumatology of his own.

129. This subcerebral space was a putative passage for air as well as for cerebral excesses (see note 80).

130. A central feature of Descartes's theory of brain function is the differential flow of spirits that leave the pineal gland to move through the ventricles and enter the pores of the ventricular lining. During the waking state, a continuous undifferentiated general flow occurs, but acts of sensation and motion require intensified local currents. There are three factors affecting these:
(a) Externally (sensorily) instigated differences in the openness and orientation of the pores themselves — since wide-open pores that face a particular point on the gland will promote a flow from that point toward those pores;
(b) Intraglandularly instigated differences in the intensity of flow from different points on the surface of the gland — since the resultant currents will force open the pores toward which the currents are directed; and
(c) Differences in the axial orientation of the gland at different times — since the degree and direction of its inclination will determine which point on the gland is opposite which pores in the lining.
As we proceed, we shall hear much about the causal factors that govern the three differentials just mentioned. One of these is the soul itself — which can voluntarily alter both the inclination of the gland and the pattern of effluence from its surface; this is the way in which the soul gives rise to voluntary movements.

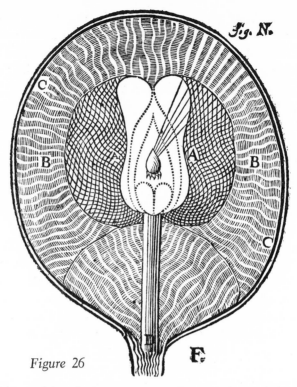

Figure 26

times this machine, being so arranged as to obey all the actions of the spirits, represents the body of a man who is *awake*. Or at the least the spirits have strength enough to push some [of the nervous filaments] in the way indicated and [thus] to stretch certain parts of the brain while others remain free and lax: as do different parts of a sail when the wind is a little too weak to fill it. And at such times this machine represents the body of a man who sleeps and who has various dreams while sleeping. Imagine, for example, that the difference between the two diagrams M and N [Figs. 27 and 28] is the same as that between the brains [a] of a man who is awake and [b] of a man who is sleeping and dreaming.[131] [Cl–69; AT–174]

131. Descartes appears to wish to link waking with general intracerebral tension or turgor, dreamless sleep with general laxness, and dreaming with local tension or turgor during a period of general laxness. There were Hippocratic, Aristotelian, and Galenic theories — variously elaborated in the Middle Ages — associating the dream state and/or dream contents with specific physiological (usually humoral) conditions (see, for example, J. Chadwick and W. N. Mann, *The Medical Works of Hippocrates*, Springfield, 1950, 194–201; Aristotle, *De somniis et vigilia*, 460b28–463b10; Galen, *De dignotione ex insomniis*, K6, 832–835). Descartes appears to derive little

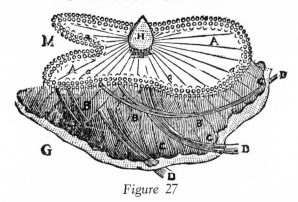

Figure 27

But before I speak to you in greater detail concerning *sleep* and *dreams*, I would have you first consider whatever is most noticeable about the brain during the time of waking: namely, how ideas of objects are formed in the place destined for *imagination* and for *common sense* [see note 72], how these ideas are preserved by *memory*, and how they cause *the movement of all the members*.

You can see in the diagram marked M [Fig. 27], that the spirits that leave gland *H*, having dilated the part of the brain marked A [the ventricle] and having partly opened all its pores, flow thence to

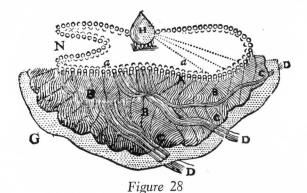

Figure 28

from these theories other than the general idea that sleep and dreaming have a physiological — for Descartes a physical — basis. He uses the slack-sail analogy elsewhere, in a letter to the Marquis of Newcastle. There he says that the turgor of the waking state keeps open the pores of the brain, permitting sensations which are no longer possible when the brain relaxes and the pores are shut down (AT 4:192). For Descartes on intraneural turgor, see also the *Dioptrics*, AT 6:111–112. For his further ideas on waking and sleeping, and for some Renaissance antecedents to his theory, see note 156.

B [the fibrous mesh of the brain substance], then to C [the membrane enveloping this mesh], and finally into D [the origins of the cranial nerves], whence they spill out into all the nerves. And by this means they keep all the filaments that compose the nerves and the brain so tense that even those actions that have barely force enough to move them are easily communicated from one of their extremities to the other, nor do the roundabout routes they follow prevent this.

But lest this circuitousness keep you from seeing clearly how this [mechanism] is used to form ideas of objects that impinge on the senses, notice in the adjacent drawing [Fig. 29] the filaments 1–2, 3–4, 5–6, and the like that compose the optic nerve and extend from the back of the eye (1, 3, 5) to the internal surface of the brain (2, 4, 6). Now assume that these threads are so arranged that if the rays that come, for example, from point A of the object happen to exert pressure on the back of the eye at point 1, they in this way pull the whole of thread 1–2 and enlarge the opening of the tubule marked 2. And similarly, the rays that come from point B enlarge the opening of tubule 4, and so with the others. Whence, just as the different ways in which these rays exert pressure on points 1, 3, and 5 trace a figure at the back of the eye corresponding to that of object ABC (as has already been said), so, evidently, the different ways in which tubules 2, 4, 6, and the like are opened by filaments 1–2, 3–4, and 5–6 must trace [a corresponding figure] on the internal surface of the brain. [Cl-72; AT-175]

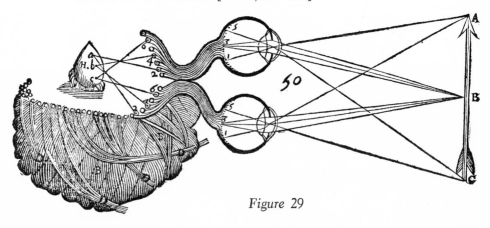

Figure 29

Suppose next that the spirits that tend to enter each of the tubules 2, 4, 6, and the like do not come indifferently from all points on the surface of gland H but each from one particular point; those that come from point a of this surface, for example, tend to enter tube 2, those from points b and c tend to enter tubes 4 and 6, and so on. As a result, at the same instant that the orifices of these tubes enlarge, the spirits begin to leave the facing surfaces of the gland more freely and rapidly than they otherwise would. And [suppose] that just as [a] the different ways in which tubes 2, 4, and 6 are opened trace on the internal surface of the brain a figure corresponding to that of object ABC, so [b] [the different ways] in which the spirits leave the points a, b, and c trace that figure on the surface of this gland.[132, 133]

And note that by "figures" I mean not only things that somehow represent the position of the edges and surfaces of objects [that is, their shape], but also everything which, as indicated above, can cause the soul to sense movement, size, distance, colors, sounds, odors, and other such qualities; and even things that can make it sense titillation, pain, hunger, thirst, joy, sadness, and other such passions. For it is easy to understand that tube 2, for example, will be differently opened by the action that I said causes a red, or titil-

132. Descartes wishes to emphasize the point-to-point correspondence among four patterns: the pattern of the object, the pattern of the retinal image, the pattern of the projection of that image on the lining of the brain cavity, and the pattern of effluence of spirits through the surface of the pineal gland. Later physiologists were to repudiate animal spirits and substitute neuronal pathways within the brain for the currents envisioned by Descartes. But the concept of a correspondence between retinal and cerebral patterns was to be episodically reaffirmed and was ultimately to become a stable part of the theory of visual perception. See also Rothschuh, D., 108.

133. In a rather oblique way, Descartes handles here a crucial aspect of his whole psychological theory: he suggests the physical basis of conscious sensation (see also notes 62 and 72 above, and Rothschuh, D., 109, note 1). An act of sensory awareness entails, in his belief, a more than ordinarily abundant outflow of animal spirits from the pineal gland toward and into particular apertures in the ventricular lining. More than once, Descartes emphasizes that the brain movements perceived directly by the soul, although they vary according to the qualities of the object perceived, are in no sense identical with these qualities (see the following paragraph and the beginning of the Sixth Discourse of the Dioptrics). The sensation we experience depends upon the nature and place of the outflow of spirits from the surface of the gland, which in turn is determined by the size and orientation of the apertures in the ventricular lining into which those spirits are to flow. Which apertures are open at a given moment, and how each is oriented, are matters determined by mechanical dislocations occurring at the peripheral ends of particular nerves.

lating, sensation than by the [action] that I said causes a white, or painful, sensation; and the spirits that leave from point *a* will tend differently toward this tube according as it is differently open, and so with other differences as well.[134] [Cl–73; AT–176]

Now among these figures, it is not those imprinted on the organs of external sense, or on the internal surface of the brain, but only those traced in spirits on the surface of gland *H, where the seat of imagination and common sense is,* that should be taken to be ideas, that is to say, to be the forms or images that the rational soul will consider directly when, being united to this machine, it will imagine or will sense any object.[135]

134. Descartes seems willing to reduce all intermodal and intramodal discrimina-tions to differences in the size and orientation of the openings through which spirits pass from the brain ventricles into the interfilamentous interstices of the brain substance. The size and orientation of these openings, and hence the ease with which spirits enter them, determine the flow pattern of spirits through the ventricle right back to their source in the pineal gland. The resultant differences in the pattern of effluence from the gland form the basis of conscious sensory distinctions (see note 133).

135. *The soul and the pineal gland.* Descartes's ideas on this subject have been adversely criticized but are not unreasonable considered in context.

This is the first passage locating psychic activity in the pineal gland. When we study the climate of ideas within which Descartes developed this notion, it seems less gratuitous than it otherwise might. There were, in general, two theories about the use of this gland: first, that it controls (in ways that differed according to different theorists) the flow of spirits between the third brain ventricle and the fourth (for Galen on this, see Siegel, *Galen's System*, 119); and second, that it supports and keeps separate the vessels that enter the brain to form the choroid plexus. Galen said that "as often as nature subdivides a raised up vessel, she interposes a gland to fill the interval" (*UP*, bk. 6, chap. 4, K3, 424).

Descartes's idea that animal spirits move from blood to gland to ventricle is a fairly natural development of the Galenic view re-expressed by Paré who said that the pineal's "utility is to reinforce the separateness [division] of the vessels brought there by a flap of the pia mater for generation of animal spirits and to give life and nourishment to the brain" (*Oeuvres*, 1575, 127). If there is to be a separate soul, what more plausible locus of intermediation between it and the body than the gland whose associated blood vessels form the chief source of animal spirits?

In pre-Cartesian Renaissance thought, circa 1542 to 1632, the functions of the pineal body were variously represented. For Columbus, "the use of this gland is for separating the veins even though the matter appears otherwise to certain anatomists to whom it seems evident that it was made to shut in the spirits of the fourth ventricle, but this idea impresses me as entirely improbable" (*Anatomica*, 192). Piccolhomini supposed that the "pineal is provided to keep open the interventricular foramen lest the brain's diastole prevent a proper backflow of spirits from the fourth ventricle to the third" (*Praelectiones*, 255). According to du Laurens, the third ventricle produces two channels, one communicating with the pituitary gland, the other with the fourth ventricle. In the latter channel appears a pointed glandule shaped like a pine cone, "deemed to serve, as other glands do, for steadying the veins

And note that I say "will imagine or will sense" inasmuch as I wish to include under the designation *Idea* all impressions that spirits receive in leaving gland *H*; and these [*a*] are all to be attributed to the common sense when they depend on the presence of objects, but [*b*] can also proceed from several other causes, as I shall later explain, and should then be attributed to imagination.[136]

And I could add something here about how the traces of these ideas pass through the arteries toward the heart and thus radiate through all the blood; and about how they can sometimes even be caused, by certain actions of the mother, to be imprinted on the limbs of the child being formed in her entrails. But I shall content myself with telling you more about how they are imprinted in the internal part of the brain, marked *B*, which is the seat of *Memory*. [Cl-74; AT-177]

With this end in view, imagine that after leaving gland *H* [see Fig. 29] spirits pass through tubes *2, 4, 6,* and the like, and into the pores or intervals that occur between the filaments composing part

and arteries which are distributed to the brain so that the animal spirits have a free and open way to go from the third ventricle to the fourth" (*Oeuvres*, 1621, 306).

A warning was sounded by Bauhin, who admonished that being external rather than internal to the brain, the pineal could not act as a sort of guard controlling the flow of spirits from the third to the fourth ventricle (*Theatrum*, 1605, 598–599). But Crooke allowed the gland *both* functions, that is, to "confirme the divisions" of the vessels that give rise to most of the blood plexuses of the brain, and "to keep the passage of the third ventricle open — so that the animal spirit [be not] hindered from descending into the fourth ventricle" (*Mikrokosmographia*, 1631, 467–468).

As between these variously interpreted and often synthesized ideas of the purpose of the pineal gland, Descartes was primarily influenced by its alleged function of sustaining the subdivisions of the vessels entering the brain. The supposition that it played this role made it a proper place for the separation of spirits from the blood and for a patterned transmission of these spirits to, and through, the ventricles. These uses of the gland made it, by the same token, a proper organ for intermediating between the body and the soul.

For Descartes elsewhere on the pineal gland as *siège de l'âme* or *siège du sens commun*, see the *Passions* (AT 11:351–362) and the *Dioptrics* (AT 6:129); also his letters to Meysonnier (AT 3:18–21) and Mersenne (AT 3:123, 263–265, and 362–363).

136. Ideas, as Descartes conceives of them here, are differentiations not of *res cogitans*, or soul, but of *res extensa*, or matter. In his epistemological writings and even occasionally in *Man*, he uses the term in the contrary sense, making ideas differentiations of soul. As a physiologist, he gives ideas a corporeal status. An idea is an "impression" received by animal spirits (composed, we remember, of terrestrial particles) as they leave the pineal gland. The "impression" is, in effect, a differentiated pattern of currents. Such patterns are partly attributable to the properties of objects of sensation, and partly to imagination (see also notes 137 and 140).

B [the solid part] of the brain. And [assume] that they are forceful enough to enlarge these intervals somewhat and to bend and rearrange any filaments they encounter, according to the differing modes of movement of the spirits themselves and the differing degrees of openness of the tubes into which they pass. [Assume also] that the first time they accomplish this they do so less easily and effectively here than on gland *H*, but that they accomplish it increasingly effectively in the measure that their action is stronger, or lasts longer, or is more often repeated. Which is why in such cases these patterns are no longer so easily erased, but are retained there in such a way that by means of them the ideas that existed previously on this gland can be formed again long afterward, without requiring the presence of the objects to which they correspond. And it is in this that *Memory* consists.[137] [Cl–75; AT–178]

137. *Memory.* Historically, Descartes's theory may be understood as a corpuscularized version of explanations set forth by Scholastic philosophers and Renaissance anatomists who in turn had elaborated Greek ideas. Plato and Aristotle compared imprinting to the effect of a seal on wax (Plato, *Theaetetus,* 191 D, E; Aristotle, *Mem.,* 450b1). Aristotle said also that memory belongs to the "primary faculty of sense perception," that is, to the common sense (*Mem.,* 450a1B–15).

In Renaissance thought, memory tended to be made coordinate with (rather than, as Aristotle suggested, an endowment of) the common sense. Thus Fernel tells us that the sentient soul has *three faculties:* common sense, imagination, and memory (*Physiologie,* 428). Paré refers to *five internal senses,* namely (*a*) the animal faculty, (*b*) the common sense, (*c*) the imaginative or estimative sense or fantasy, (*d*) the cogitative or ratiocinative sense or understanding, and (*e*) memory ('De la génération de l'homme,' in "*Deux livres de chirurgie,*" Paris, Wechel, 1573, 48–67, republished with some changes in *Oeuvres,* 1579, 856–858). Piccolhomini says that the sentient soul has *two cognitive faculties:* (*a*) the common sense used for discrimination, and (*b*)another faculty for imagining, reasoning, and remembering (*Praelectiones,* 244). For still further variations on this basic theme, see Bauhin (*Theatrum,* 570), Bartholin (*Enchiridion,* 809–812), and others.

The above Renaissance theories were debated in the light of several issues and questions: Is memory a sense, or a faculty? Where is its seat in the brain? Is there an intellective type of memory in addition to a sensitive type? (This duality was widely proposed.) The sealing-wax analogy recurred with considerable frequency in Renaissance psychology from the time of Juan Vives (*De anima et vita,* Basel, 1538, 54). For Scholastic antecedents of some of these ideas, see Gilson, *ISC,* 175–179.

Descartes seems, at first, little influenced by these essentially facultative (Aristotelian-Galenic) theories of memory. But in his *Rules for the Direction of Our Mental Power* (*Regulae*), he speaks of a cognitive force (*vis cognoscens*) that interacts with common sense, imagination, and memory — figuratively acting now like a seal, now like sealing wax — and says that this force itself may be said to sense, imagine, and remember (AT 10:415–416).

Memory entails, in Descartes's opinion, a fifth pattern in addition to the four he has already mentioned in connection with sensation. That is, to the patterns of (*a*) *the object,* (*b*) *the retinal image,* (*c*) *the brain lining,* and (*d*) *the spirits leaving*

For example, when the action of the object *ABC*, by enlarging the degree of openness of the tubes *2*, *4*, and *6*, causes the spirits to enter therein in greater quantity than they otherwise would, it gives these spirits force enough, as they pass on farther toward *N*, to form certain passageways there [no letter *N* used to represent indicated part of brain substance]. These passageways remain open even after the action of object *ABC* has ceased; or at least, if they close again, they leave a certain arrangement of the filaments composing this part of brain *N* by means of which they can be opened more easily later than if they had not been opened before. Similarly, if one were to pass several needles or engravers' points through a linen cloth as you see in the cloth marked *A* [see Fig. 30], the little holes that one would make would stay open as at *a* and at *b* after the needles had been withdrawn; or if they closed again, they would leave traces in this cloth, as at *c* and at *d*, which would enable them to open quite easily again.

---

the gland is now added (*e*) a pattern of *openness in the intervals* between the fibers of the brain substance and, associated therewith, a pattern of *flexures* in the fibers themselves. Repeated recall reinforces this fifth pattern, he believes, thus making it easier each time for spirits to flow into the spatial intervals thus established. For more on this point, see the *Passions*, AT 11:360.

In letters to Mersenne (AT 3:19–20 and 48), Descartes compares the storing of something in memory to creasing a paper, and thereafter frequently uses the term "memory creases" (*plis de mémoire*). He thinks that "the species [a term analogous to Peripatetic 'forms'] that serve memory" may be located in the conarium, but that they are found in other parts of the brain as well as in nerves, muscles (the lute player's fingers have memory), and even in things outside the body. There is, however, an additional sort of memory that is purely intellectual and belongs exclusively to the soul (letters to Mersenne, AT 3:84–85 and 143; to Huyghens AT 3:580; and to Mesland AT 4:114).

As to memory creases, Descartes acknowledges the possibility of their interfering with one another, so that their numbers, though great, must be limited. But separate creases are not needed for each remembered item, one crease sufficing to remind the soul of a whole class of similar ones. He rejects the notion that we best remember the things that happen earliest in life; re-remembrance (reinforcement) is more decisive in this respect (AT 3:83 and 143–144). Intellectual memory has entirely separate "species" which in no wise depend on the physical creases of the brain (letter to Mesland, AT 4:114).

For Descartes elsewhere on corporeal vs intellectual memory, see the references in *ISC*, 175 and in the subject index to the *Oeuvres*, AT suppl., 85. For a possible Scholastic source on the distinction between the two sorts of memory, see Gilson, *ISC*, 176, "memorandi facultas duplex est; una sensitiva . . . altera intellectiva" (Coimbran commentators on Aristotle). For Descartes on memory, see also Rothschuh, D., 111–112, note 2. Leads to many Renaissance ideas on this subject are to be found in M. N. Young, M.D., *Bibliography of Memory*, Philadelphia and New York, Chilton, 1961.

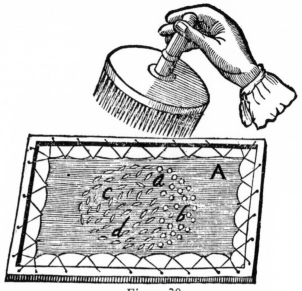

Figure 30

And it is similarly necessary to remark that if one were merely to reopen some, like *a* and *b*, that fact alone could cause others like *c* and *d* to reopen at the very same time, especially if they all had been opened several times together and had not customarily been opened separately. Which shows how the recollection of one thing can be excited by that of another which was imprinted in the memory at the same time. For example, if I see two eyes with a nose, I at once imagine a forehead and a mouth and all the other parts of a face, because I am unaccustomed to seeing the former without the latter; and seeing fire, I am reminded of heat, because I have felt the latter in the past when seeing the former.[138] [CI–77; AT–179]

138. *Associative memory.* Plato said, in this connection, that the sight of the lyre may evoke the memory of the lyre player as the portaitist evokes the sitter — or the sitter's friend. Aristotle developed a theory of willed recollection based on kinetic sequences; in an effort to recollect experience *M*, you think of some other experience (say *K*) that can lead you in kinetic sequence (through *L*) to *M*. The sixteenth and early seventeenth centuries saw an outpouring of works on memory, in some of which association was considered in its theoretical or practical aspects; Descartes is not specific enough to enable us to identify the sources of his thought on this subject. We may agree with Rothschuh (*D.*, 112, note 1) that Descartes's "mechanism for explaining association is not very convincing." For medieval and Renaissance works on mnemonics, see M. N. Young, *Bibliography of Memory*, note 137 above, and F. A. Yates, *The Art of Memory*, London, Routledge and K. Paul, 1966.

Consider furthermore that gland *H* is composed of matter which is very soft and that it is not completely joined and united to the substance of the brain but only attached to certain little arteries whose membranes are rather lax and pliant, and that it is sustained as if in balance by the force of the blood which the heat of the heart pushes thither. [And suppose that] therefore very little [force] is required to cause it to incline and to lean, now more now less, now to this side now to that, and so to dispose the spirits that leave and make their way toward certain regions of the brain rather than toward others.

Now there are two principal causes (not counting the force of the soul of which I shall speak later on) which can make the gland move in this way, and I must explain these here.

First are the differences found among the particles of the spirits that leave it. For were these spirits all of exactly equal force and if no other cause made the gland lean this way or that, then the particles would flow equally in all its pores and keep it erect and immobile at the center of the head, as is represented in diagram number 40 [Fig. 31]. But just as a body attached only by threads and sustained in the air by the force of fumes leaving a furnace would incessantly float here and there as the different particles of the fumes acted differently against it, so the particles of the spirits that hold up and sustain this gland, almost always differing among themselves in some way, do not fail to agitate it and make it lean now to one side and now to the other. See thus in diagram 41 [Fig. 32] that not only is the center of gland *H* a slight distance away from the center of the brain (marked *o*), but also the extremities of the arteries that sustain it are curved in such a way that nearly all the spirits that these arteries bring there proceed through region *abc* of its surface toward the tubules 2, 4, and 6 — in this way

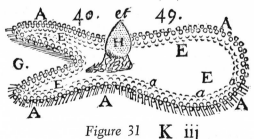

Figure 31   **K iij**

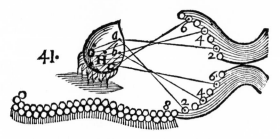

Figure 32

opening pores that face in that direction much more than they do other pores.[139] [Cl–78; AT–180]

Now the chief effect that follows from this is that the spirits, departing from certain regions on the surface of this gland and not from others, have force enough [to do two things]. [1] They can turn the tubules into which they flow, in the inner brain surface, toward the places where these spirits emanate from the gland (unless the tubules in question are already pointed in that direction). And [2] they can make the members to which these tubules correspond turn toward places corresponding to the indicated regions on the surface of gland *H.* And note that if we have an idea about moving a member, that idea — consisting of nothing but the way in which spirits flow from the gland — is the cause of the movement itself.[140] [Cl–79; AT–181]

[In Fig. 33,] for example, one can suppose that what makes tube 8 turn toward point *b* rather than toward some other point is merely that the spirits leaving point *b* tend with greater force toward 8 than do any other [spirits].[141] The same thing will cause the soul to

139. To produce a particular motor response, the gland must direct spirit particles from the proper part of itself into the proper nerve tubules; and hence it must lean this way or that. Which way it leans depends on (*a*) the differential forces of spirits leaving the gland, and (*b*) the curvatures of the arteries by which the gland is held up. The key to the model is that, for a muscle to turn a part (an arm, for instance, or the eyes) in a certain direction, a particular point on the surface of the gland must be opposite the orifice of that muscle's nerve tubule. For Descartes further on this, see the *Passions,* AT 11:361–362; also Rothschuh, D., 114, note 1.

140. Earlier, we heard Descartes equate ideas with patterned flows of spirits, not with differentiations of thought (see note 136). Ideas, thus materially defined, are involved even in involuntary movements of which the soul is not a cause. In certain ways, the heart of Descartes's endeavor, after he had split reality into material and mental components, was to ascribe as much as possible — ideas included — to the former, that is, to *res extensa* rather than *res cogitans.*

141. We may ask how different orientations of the orifice of a given nerve in the brain can differently influence the activity of that nerve at its outer end. Descartes does not give a crystal-clear answer to this question.

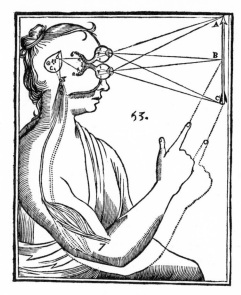

Figure 33

sense that the arm is turned toward object B provided the soul is already in this machine, as I shall later suppose it to be. For it is necessary to think that all points of the gland toward which tube 8 can be turned correspond to places toward which the arm marked 7 can be turned. Thus what makes the arm turn toward object B, at a particular time, is simply that at that time the tube for this arm is facing point b of the gland.[142] But if the spirits, changing their course, were to turn this tube toward some other point of the gland, say toward c, then filaments 8 and 7 arising nearby and proceeding to the muscle of this arm, by thus changing their position, would close certain of the pores of the brain which are near D and would enlarge others. This would make the spirits, passing thence into these muscles otherwise than they now do, promptly turn this arm toward object C. Reciprocally, if some action other than that of the spirits which enter tube 8 were to turn this same arm toward B or toward C, this would make tube 8 turn toward points b or c of the gland. As a result, an idea of this movement would be formed

142. Here and elsewhere (see notes 107 and 108) Descartes indicates that the soul knows the relative degrees of contraction of various muscles. He views this knowledge as resulting from the soul's awareness of differential flow patterns from the pineal gland to various nerve orifices. Today we ascribe such information to kinesthetic nerve impulses.

at the same time, at least if one's attention were not diverted — that is to say, if gland *H* were not prevented from leaning toward 8 by some other action which was stronger. Thus in general one must suppose [*a*] that each tubule in the internal surface of the brain corresponds *to a member*, and [*b*] that each point of the surface of gland *H* corresponds *to a direction* toward which these members can be turned: whence the movements of these members, and the ideas thereof, can be reciprocally caused the one by the other. [Cl–81; AT–182]

Furthermore, when the two eyes of this machine (and the organs of the several other senses) are directed toward one and the same object, there are formed not several ideas of it in the brain, but only one. To understand this, one must suppose that [in such circumstances] spirits leaving the same points on the surface of gland *H* are able — by tending toward different tubes — to turn different members toward the same objects. Thus [in Fig. 33] spirits leaving the same point *b* — by tending toward tubes *4, 4*, and *8* — simultaneously turn the two eyes and right arm toward object *B*.[143] [Cl–82; AT–183]

This you will easily believe by using it to grasp what the idea of distance consists in. Assume that as the gland is altered in position, the closer a point on its surface is to *o* (the center of the brain) the more distant is the place thus referred to, and that the farther the point is from *o* the nearer that place is. In the present case, for example, one assumes that should point *b* be pulled somewhat farther to the rear than it is, it would correspond to a place more distant than *B*; and if it were made to lean a little farther forward, it would correspond to a place that was nearer.[144]

143. Descartes here considers the collation of different perceptions. If a group of sense organs are muscularly trained on one object, the central ends of the nerves of the muscles involved will be oriented toward — and will tend to receive impulses from — a common point on the gland. Thus a single idea of the object is formed, "idea" in this instance being defined as a patterned efflux of spirits from the gland (see notes 136 and 140).

144. Earlier, we have heard that the distance of objects is signaled by such things as (*a*) the degree of flatness of the lens, (*b*) the muscular strains that accompany convergence, (*c*) past experience permitting inference from size, and (*d*) distinctness of outline or of color. Now Descartes suggests, in addition to these cues, another — namely the degree of deflection of the pineal gland and correlated flow patterns of spirits.

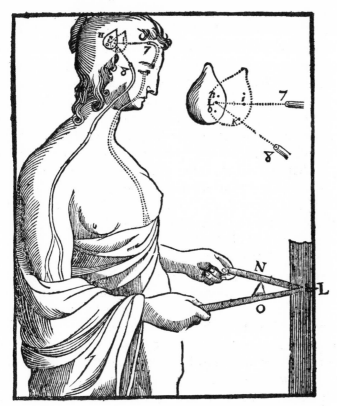

Figure 34

And this will permit the soul, when there will be one in this machine, to sense different objects through the mediation of the same organs similarly arranged, no change occurring other than in the situation of gland *H*. For example [see Fig. 34], the soul will be able to sense what is at point *L* through the mediation of the two hands holding sticks *NL* and *OL*. For spirits that enter tubes 7 and those that enter tube 8 (to which the two hands correspond) both leave from the same point *L* of the gland. But now suppose that gland *H* were leaning a little farther forward, so that the points *n* and *o* on its surface were at the places marked *i* and *k*. Now the spirits entering 7 and 8 would leave the gland at points *i* and *k*. The soul would be able to sense both what is at *N* and what is at *O*, and would do so by the mediation of the same hands without their having been in any way changed. [Cl–83; AT–184]

It remains to be noted that when gland H is inclined in one direction by the force of the spirits alone, without the aid of the rational soul or of the external senses, the ideas that are formed on its surface proceed not only [a] from inequalities in the particles of the spirits causing corresponding differences in temper, as mentioned before [see note 122], but also [b] from the imprints of memory. For if at the region of the brain toward which the gland is inclined, the shape of one particular object is imprinted more distinctly than that of any other, the spirits tending to that region cannot fail to receive an impression thereof. And it is thus that past things sometimes return to thought as if by chance and without the memory of them being excited by any object impinging on the senses.[145] [Cl–84; AT–184]

But if several different figures are traced in this same region of the brain almost equally perfectly, as usually happens, the spirits will acquire a [combined] impression of them all, this happening to a greater or lesser degree according to the ways in which parts of the figures fit together. It is thus that chimeras and hypogryphs are formed in the imaginations of those who daydream, that is to say who let their fancy wander listlessly here and there without external objects diverting it and without the fancy's being directed by reason.

But the effect of memory that seems to me to be most worthy of consideration here consists in [the fact] that without there being any soul in this machine it can be naturally disposed to imitate all the movements that real men (or many other, similar machines) will make when the soul is present. [Cl–85; AT–185]

The second cause which can determine the movements of gland H is the action of objects that impinge on the senses. For it is easy to understand [see Fig. 35] that when the degree of openness of tubules 2, 4, and 6, for example, is increased by the action of object

---

145. Ideas, we remember, are the differential flow patterns assumed by spirits leaving the pineal gland (see note 136). Four things can form or influence these patterns: (a) activities of the soul (not thus far considered by Descartes), (b) newly arriving sense perceptions (represented by reorientations of tubule openings in the ventricular lining), (c) inequalities among the particles of the spirits themselves, and (d), we now hear, memory. Memory traces, we were earlier told, consist in residual patterns of openness among the interstices of the filamentous brain substance. When the pineal gland accidentally leans in such a way that flow patterns are influenced by memory traces, unsolicited recollections are evoked.

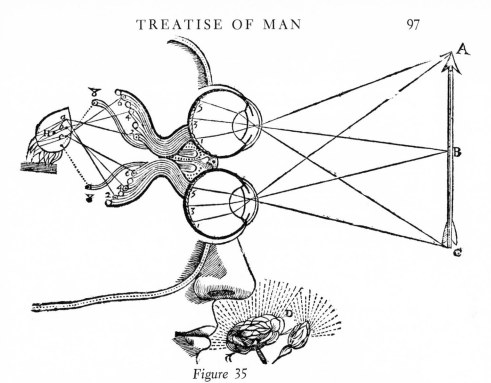

Figure 35

ABC, the spirits, which commence at once to flow toward them
more freely and rapidly than they did [before], draw the gland after
themselves a little, and cause it to lean if it is not otherwise prevented from so doing; so that, changing the position of its pores,
it begins to conduct a much greater quantity of spirits through *a, b,*
and *c* to 2, 4, and 6 than it did before: which renders the idea that
these spirits form correspondingly more perfect. This constitutes the
first effect that I wish you to notice. [Cl–86; AT–185]

The second consists in [the fact] that while leaning thus to one
side this gland is prevented from easily receiving ideas of objects acting on other sense organs. For example, [see Fig. 35], during the
time when almost all the spirits that gland *H* produces leave it from
points *a, b,* and *c,* too few leave from point *d* to form there the idea
of object *D* whose action I assume to be neither as lively nor as
strong as that of object *ABC.* Whence you see how ideas mutually
impede one another and why one cannot be strongly attentive to
several things at one time.

It is also necessary to note that when sense organs are first impinged upon by a particular object more than by others, but are

not yet maximally disposed to receive its action, the mere presence of the object suffices to dispose them thereto completely. Thus, for example, if the eye is disposed to look at a very distant place at the moment when a very near object *ABC* first presents itself, I say that the action of this object will be able at the same instant to redispose the eye in such a manner as to fixate the object.[146]

And in order that you may be able to understand this matter more easily, consider first the difference that exists between an eye disposed to look at a faraway object as in diagram 50 [Fig. 29] and the same eye arranged to look at a nearer object as in diagram 51 [Fig. 36]. The difference consists not only in the facts

[a] that the crystalline humor (in the latter case) is a little more arched; and

[b] that other parts of the eye [for example the pupil, see notes 104 and 105] are, correspondingly, differently disposed; but also in the facts

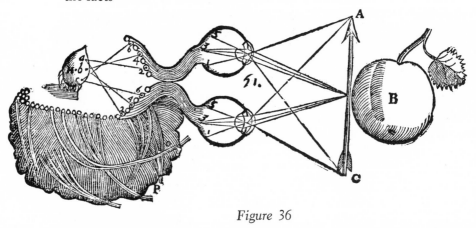

Figure 36

146. Descartes here stresses more cogently than earlier theorists the fact that vision is ordinarily maximal with respect to one particular object point at a time, and rightly supposes that this fact entails a constellation of muscle reflexes (to be called, by later theorists, fixation). The fixating reflexes, he realizes, both alter the shape of the lens and reorient the eye as a whole. The central mechanism must be one that can proceed entirely automatically or, under certain conditions, with conscious intervention. The logical vehicle for this sort of response, given Descartes's model of the brain, would be the pineal gland. However unsatisfactory his model may seem from today's perspective, he deserves some credit for realizing that a cerebral mechanism for reflex control is needed. In the succeeding paragraphs, he gives the specifics of this mechanism as he views it. Rothschuh correctly notes (D., 122, note 1) that fixation involves for Descartes a cybernetic aspect, a kind of intensification of fixation through positive feedback (*Rückkoplung*).

[c] that tubules 2, 4, and 6 are inclined [so as to orient the eye] toward a nearer [object] point;

[d] that gland H is a little more advanced toward these tubules; and

[e] that the region a, b, c of the gland's surface is proportionately a little more curved or arched so that, in both figures, spirits from point a always tend toward tube 2; from b, toward tube 4; and from c, toward tube 6. [Cl–87; AT–186]

Consider also that the movements of gland H are sufficient in themselves to change the position of these tubes and as a result [to change] the disposition of the eye as a whole, the general point having just been made that these tubes can make all members move. [Cl–88; AT–187]

Consider next that the tubes 2, 4, and 6 [see Fig. 36] can be opened by the action of object ABC to the same extent that the eye is disposed to look at that object. If, for example, the rays that fall on point 3 all come from point B, as they do when the eye looks fixedly at B, their actions will evidently pull filament 3–4 more strongly than if they came some from A, some from B, and some from C as they would do if the eye were slightly differently disposed. In the latter case, their actions, being neither as similar nor as united as before, cannot be as strong; often they even impede one another. But this [that is, clear vision] occurs only in the case of objects whose outlines are neither too uniform nor too confused, as is true also of objects whose shape and distance the eye can distinguish, as I have remarked in the *Dioptrics*.

Consider further that gland H can be moved in the direction that increases — more easily than in that which decreases — the eye's disposition to receive the action of whatever object is already acting most strongly upon it. If, in diagram 50 [Fig. 29] for example, we suppose ABC to be a near object acting strongly on an eye disposed for distant vision, less force is required to incite the gland to lean forward than backward. For in leaning backward it would render the eye less oriented than before to receive the action of object ABC (which we suppose to be a near object and to be the one acting on the eye most strongly). And it would cause the tubules 2, 4, and 6 to be, to that extent, less opened by the action of the object; and spirits leaving the gland at a, b, and c would flow, to the same extent, less freely to these tubules. Whereas, in leaning forward, it would better dispose the eye to receive the action of the object; tubules 2, 4, and 6 would open more; and, as a result, spirits leaving the gland at *a, b,* and *c* would flow toward the tubules more freely. [And this occurs] in such a way that, as soon as the gland begins to move to even the slightest degree, it is borne at once on the current of the spirits and is not permitted to stop until it is entirely disposed in the manner that you see in diagram 51 [Fig. 36] where the eye is looking fixedly at nearby object *ABC.* [CL–89; AT–188]

It thus only remains for me to tell you what cause can thus initiate the movement of the gland. This is ordinarily nothing but the force of the object itself which, acting on a sense organ, augments the openness of certain of the tubules in the internal surface of the brain, so that spirits, beginning promptly to flow toward these tubules, draw the gland with them and make it lean in that direction. But in case the tubes on this side have already been otherwise opened to the same or a greater extent than this object would open them, we must suppose that the spirit particles flowing out in every direction through the pores of the gland, being unequal, will push the gland hither and thither in the twinkling of an eye without giving it a moment's respite. And if they should happen first to push it in a direction in which it is not easily inclined [because the pores on that side are not open very wide] their action, being in itself rather slight, will have very little effect. But as soon as they push it ever so slightly in the direction in which it is already carried [by the already widely opened pores], it will not fail to lean that way promptly and, as a result, to dispose the sense organ to receive the action of its object in the most perfect possible way, as just explained. [Cl–91; AT–189]

Let us have done with conducting the spirits to the nerves, and see now what movements depend on them. Suppose that none of the tubules of the internal surface of the brain are more open than any others, and that consequently the spirits have no impression in them of any particular idea [see note 136]. [In such a case] they [the spirits] will spread indifferently in all directions and pass from the conduits which are near *B* [the deeper regions of the brain mass] and into those near *C* [the superficial regions of the brain mass], whence the most subtle of their parts flow off directly from the brain through the pores of the little membrane which envelopes it [the pia mater], while the rest, making their way toward D, will proceed into the nerves, and thence into the muscles, without causing there any particular effect because they will be distributed to all muscles equally. [Cl–92; AT–189]

But if by the action of the objects that move our senses, some tubules are more or less opened — or even just differently opened — than others, the filaments that compose the brain substance being consequently some a little more tense or more slack than others, will conduct the spirits toward certain regions near its base and thence toward certain nerves with more or less force than toward others. Which will suffice to cause different movements in the muscles, in accordance with what has already been amply explained.

Now I would have you conceive of these movements as similar to those to which we men are incited by the different actions of objects impinging on our senses. To this end I would have you consider six different conditions to which the different movements may be due. The first is the place of origin of the action that opens the tubules which the spirits first enter in the brain. The second consists in the force and all the other qualities of this action. The third [consists] in the arrangement of the little threads that compose the substance of the brain. The fourth, in the unequal force that the particles of the spirits can have. The fifth, in the different positions of the external members. And the sixth, in the interplay of many actions that move the senses simultaneously.

As to the place whence the action proceeds, you already know that if the object ABC [Fig. 36], for example, were to act on some sense other than vision, it would open other tubes in the internal surface of the brain than those marked 2, 4, and 6. And [you already know] that if it were nearer or farther away or otherwise situated with respect to the eye than it is, it could in truth open the same tubes but they would have to be otherwise situated than they are, and would be able therefore to receive spirits from other points of the gland than those which are marked a, b, c and to conduct them toward other regions than ABC where they conduct them now, and so in other cases.[147] [Cl–93; AT–191]

147. Descartes begins here to develop a more detailed theory of automatic responses than he has presented thus far. In order to account for the ability of his machine to respond in different ways to differently located objects, he makes use of the principle

As for the divers qualities of the action that opens these tubes, you know also that according as these qualities differ they open the tubes in different ways, and one must suppose that this fact alone suffices to change the course of spirits through the brain. For example, if object ABC [in Fig. 36] is red, that is, if it acts on eye 1, 3, 5 in the manner requisite, as already mentioned, to cause the color red to be sensed, and if the object has in addition the shape of an apple or other fruit, one must think that it will open the tubes 2, 4, and 6 in a particular way which will cause the parts of the brain near N to press against one another a little more than they usually do with the result that spirits entering tubes 2, 4, and 6 will make their way from N through O toward P. Whereas if object ABC were of another color or shape, it would not be the filaments near N and O that would deflect the spirits entering 2, 4, and 6 but certain other, neighboring ones.[148] [Cl–94; AT–191]

And [in Fig. 37] if the heat of the fire A, which is close to hand B, were only moderate, one must suppose that the way it opened tubes 7 would cause parts of the brain near N to be pressed together and those near O to be spread apart a little more than they usually are. Thus the spirits that come from tube 7 would go from N through O toward P. But if one supposes that this fire burns the hand, one must think that its action opens tubes 7 so wide that the spirits that enter there are strong enough to pass farther, in a direct line, than merely to N; namely as far as O and R where, pushing before them the parts of the brain that they find in their way, they are

---

that for a specific response to be given, specific sensory information must be received. He says that differently located objects open different tubes — or even the same tubes "differently situated" (*ces mesmes tuyaux . . . autrement situés*). The meaning of the latter phrase is obscure, but his central point is clear, namely that for differently located objects spirits will leave the gland at different points on its surface. When Descartes says that for objects other than ABC the spirits will no longer be conducted "toward ABC," he perhaps means that they will no longer be so conducted as to direct the eye toward ABC.

148. Descartes here attempts to correlate the qualities of sense objects and the kinds of brain responses evoked. When he says that different qualities open the tubules in "different ways," he means that the recipient orifices in the brain lining can be diversely oriented and can be opened in differing degrees. There will be corresponding differences in the pattern of spirits entering these orifices and consequently in the arrangement of the filaments of the brain substance. In what follows, he will emphasize that the differential displacement of these filaments determines which nerves will receive the spirits and hence which muscles will respond.

Figure 37

resisted by these parts and deflected toward *S*, and so in other [cases as well].[149]

As to the arrangement of the filaments that compose the brain substance, it is either acquired or natural [innate]; and since acquired [arrangements] depend on circumstances that change the

149. Descartes makes a transition here from the fixation problem to the problem of reflex responses in general, showing how "actions" — we should term them stimuli — that differ qualitatively and/or quantitatively give rise to correspondingly different physical events in the brain. However repellent the explanation in its specifics, it is significant in its recognition that different stimuli produce different responses

flow of the spirits, I shall be better able to explain them later on.
[First, then,] to explain what the natural [arrangements] consist in.
Know that in forming the filaments God arranged them as follows.
The passages that He left among them are able to conduct the
spirits, when these are moved by a particular [stimulant] action,
toward nerves that permit just those movements in this machine
that a similar action could incite in us when we act through natural
instincts.[150] Thus, [in Fig. 37] for example, if fire A burns hand B
and causes the spirits entering tube 7 to tend toward O, these spirits
find there two pores or principal passages OR and OS. One of
them, namely OR, conducts the spirits into all nerves that serve to
move the external members in the manner necessary to avoid the
force of this action, such as those that withdraw the hand or the
arm or the entire body and those that turn the head and the eyes
toward this fire in order to see more particularly what must be done
for protection against it. And through the other passage, OS, the
spirits enter all those nerves that cause internal emotions like those

---

and that a physical basis for such correlations must exist in the brain. Today we
tend to think of cellular circuitry as the mechanism involved, but the molecular events
responsible for the circuitry are still unknown.

150. Ancient and medieval authors acknowledged that the body contains both a
predisposition and an ability to give complex automatic (unlearned) responses, that
is, to behave in a manner which would later be termed "instinctive." The following
are two examples from authors known to have influenced Descartes extensively in
his views on this subject.

Galen said that "not only did she [Nature] prepare a mouth, esophagus, and
stomach, as instruments of nutrition; she also produced an animal that understands
right from the beginning how these are to be used, and she instilled into it a certain
instinctive faculty of wisdom by which each animal arrives at the nutriment suitable
for it" (May, *Galen*, vol. 2, 673). And Aquinas said that an instinct of nature moves
lower animals to specific activity determined by the forms received by the senses,
without any reasoned awareness of the end in view (*Summa Theologica*, part 1,
question 18, article 3). The problem of what we term instinctive behavior therefore had already been posed for Descartes by these and other earlier thinkers. His
task, he believed, was to supply a mechanical — specifically a *micromechanical* —
explanation of such behavior (see his letter to Newcastle, AT 4:573–576).

Commenting on Descartes on the subject of instinct, La Forge says: "Now by
instinct, and by the actions attributable thereto, be it in man or beast, our author
merely means . . . that secret disposition of the invisible parts of the body of the animal, and principally of its brain, according to which, after being imprinted by an
object, man feels incited and inclined — and the animal feels compelled — to make
appropriate actions and movements. I [La Forge] say that man is 'incited and inclined' because the Soul, despite the disposition of the body, can prevent these
movements when it has the ability to reflect on its actions and when the body is
able to obey" (*l'Homme*, 379).

that pain occasions in us, such as nerves that constrict the heart, agitate the liver, and other such.[151] Through OS they also enter nerves that can cause external movements testifying [to the internal emotions], those for example that provoke tears, or that wrinkle the forehead and cheeks, or that dispose the voice to cry. Whereas if hand B were rather cold, and fire A were to warm it moderately without burning it, that would cause the same spirits that enter tube 7 to proceed no longer to O and to R, but to O and to P where again they would find pores arranged to conduct them into all the nerves that can serve for movements appropriate to this [particular stimulant].[152] [Cl–96; AT–193]

151. *Emotion.* The term *émotion,* derivative from the Latin *emovere* (to excite or disturb), already appears in Old French in more or less its modern sense. The nature of "emotions," a term used somewhat interchangeably with "passions," was argued at length in Scholastic thought, and Descartes's *Passions of the Soul* is in part his contribution to that debate. The history of theories of emotion is too long to be given here even in outline, except to suggest that Descartes's approach to the subject had two ends in view: one definitional and taxonomic, the other analytic. Let us look briefly at both aspects of his thinking.

(*a*) Emotion had been made, in medieval thought, an intense or exacerbated activity of the soul, and following a neo-Platonic tradition, some Schoolmen spoke of all such activity as being either irascible or concupiscent (see Gilson, *ISC*, 209–210). Descartes agrees that passions are activities of the soul but disagrees that they are reducible to two. In the *Passions* he acknowledges six primitive ones: wonder, love, hate, desire, joy, and sadness (AT 11:380). Elsewhere he admits only four, and refers to these as innate: joy, love, sadness, and hate (see note 116 above). Whatever the number — four or six — he is explicit in his opposition to the idea of only two.

There were others who anticipated Descartes in revising the Scholastic idea of two primitive emotions, among them Juan Vives ("De anima" [first publ. Basel, 1538] in *Opera omnia,* Valencia, 1782, 426–520) and Lelius Peregrinus (*De affectionibus animi noscendis,* Rome, 1598). Their influence on Descartes is problematical. For pre-Cartesian theories of emotion, see A. Levi, *French Moralists: The Theory of the Passions,* Oxford, Clarendon, 1964.

(*b*) Here and in the *Passions* Descartes suggests a mechanical — that is, a corpuscular — explanation of emotion (AT 11:379–380). A similar proposal appears in a letter to Elizabeth, in which he argues that merely to contrast passion with action is to distinguish inadequately between passion, sensation, imagination, and temperament. He proposes that it is better to think of passions as occasioned by special agitations of spirits — which, we remember, are third-element (earthy or terrestrial) particles in his theory (AT 4:310–311).

In the *Principles,* he again approaches the problem in a reductive manner and says that joy depends on nerves to the heart, diaphragm, and so forth, and that very pure, well-tempered blood, since it dilates the heart unusually easily and forcefully, can stretch and move the heart's nerves in a manner that causes them to excite joy in the soul (AT 8:316–317 [9:311]). Rothschuh (*D.,* 127, note 1) properly observes that for Descartes, who assumes that man comprises a soul united with a body, emotion is for the soul an experience (*ein Erleiden*) and for the body a physical act (*ein Tun*).

152. Earlier (see note 60) Descartes suggested mechanical models for relatively simple reflexes. Here we find complex and intgrated ensembles of reflexes mentioned.

Note that I have expressly distinguished two pores OR and OS in order to call to your attention that two sorts of movements almost always follow every [sensory] action, namely external movements that serve in the pursuit of desirable things and the avoidance of injurious ones; and internal movements that we commonly designate *passions*. Passions serve to dispose the heart, the liver, and all the other organs that determine the temperament of blood — and consequently of spirits — in such a way that the spirits formed at a given time will be those suited for producing the external movements that follow. For suppose that the different qualities of these spirits are among the circumstances that affect their direction of flow (as I shall explain in a moment). In that case, one may easily believe that when it is a question of forcefully avoiding some evil by overcoming it or by driving it away — as anger inclines us to do — then the spirits must be more unevenly agitated and stronger than they usually are. Whereas, when it is necessary to avoid harm by hiding or by bearing that harm with patience — as fear inclines us to do — then the spirits must be less abundant and weaker. For this purpose the heart must constrict at such a time, and must husband and save the blood against need. Corresponding judgments can be made about passions other than these.[153] [Cl–98; AT–194]

As for other external movements, ones which serve neither to ward off the evil nor to pursue the good, but which merely bear witness to the passions — such movements as those of laughing and weeping — these occur only by chance and because the nerves through which spirits flow to produce them have their origins quite close to those that spirits enter in order to give rise to the passions, as anatomy can teach you.[154]

But I have not yet made you see how different qualities of the spirits can differently affect the direction of their flow. This happens chiefly when these spirits are but slightly — or not at all — directed

153. Modern theory both confirms and contradicts Descartes's idea about the physical basis of emotion. We agree that the blood is altered when an emergency must be met (blood sugar and oxygen are heightened, the hormonal balance is changed). We totally disagree with him about the nature of the changes that occur and about the causal mechanisms involved.

154. We are reminded that Descartes hoped, by anatomizing brains, to learn something about the basis of psychic activity.

by other causes. Suppose, for example, that the stomach's nerves are agitated in the way required, as mentioned earlier, to produce a sensation of hunger. But suppose also that at that time nothing properly edible presents itself to any of the senses or to memory. In such a case the spirits that will be caused by this action to enter tubes 8 in the brain [see Fig. 35] will proceed to a region where they will find many conduits so arranged as to conduct them indifferently into all nerves that can serve for the search or pursuit of some object. [In this situation] the sole possible cause of their making their way through certain nerves rather than others will be the inequalities of their part[icle]s.

And if it happens that the strongest particles tend now to flow toward certain nerves, and immediately afterward toward their opposites, that will make this machine simulate the movements seen in us when we hesitate and are in doubt about something.

Quite similarly, if the action of fire A [in Fig. 37] is intermediate between actions that can conduct the spirits toward R and those that can conduct it toward P, that is, between those causing pain and those causing pleasure, it is easy to understand that it must be the inequalities of the spirits alone that direct them to the one or the other: just as the same action [stimulus] that is agreeable to us when we are in a good humor can often displease us when we are sad and sorrowful. And from this you can deduce the reason for all that I have said heretofore concerning the humors or inclinations, both natural and acquired, that depend upon the differences among the spirits. [Cl–100; AT–195]

As to the [effect of] divers positions of external members, it is only necessary to imagine that they change the pores that carry the spirits immediately into the nerves. Suppose, for example, that fire A burns hand B, and suppose that the head be turned toward the left (instead of, as now, toward the right). Then the spirits will still go, as now, from 7 to N, thence to O, and thence to R and to S. But from R, instead of going to X (through which I suppose they must pass to hold the head erect when turned, as now, toward the right) they will go to Z (which I suppose they would have to enter to hold it erect if it were turned toward the left). For the position of the head which now makes the filaments of the brain substance near X more slack and more easily separable than those near Z will, if it is changed [from right to left], make those at Z very slack and those at X tense and tight.

Similarly, to understand how a single action can, without changing, move first one foot of this machine and then the other, as is required for it to walk, it suffices to suppose that the spirits pass through a single conduit [in the brain lining], a conduit whose extremity [that is, whose inner orifice] is differently disposed — and so conducts them into different nerves — when the left foot is advanced than when the right is. And relatable to this is all that I have said hitherto concerning respiration and such other movements as do not ordinarily depend on any idea [that is, reflexes]; I say ordinarily, because they can also depend upon them at times.[155] [Cl-101; AT-197]

I believe I have sufficiently explained all the functions of the waking condition, and there only remain a few things to say about *sleep*. Firstly, one need but glance at diagram 50 [Fig. 38], noting how slack and pressed together are the filaments D,D that enter the nerves, to understand how, when this machine corresponds to the body of a person asleep, the actions of external objects are mostly kept from reaching the brain (where they must go in order to be sensed); and the spirits in the brain are kept from reaching the external members (where they must go in order to move them). These are the two chief aspects of sleep. [Cl-102; AT-197]

155. The subject here is volition, and volition begins with ideas. Ideas, we recall, are currents in the spirits contained in the ventricles. They are differential currents created as the spirits enter the ventricles from the gland (see note 136). In this passage, Descartes refers specifically to currents caused by conscious command. Without thinking about the detailed mechanics, the soul determines through which pores in the surface of the gland the spirits will flow into the ventricles with special force. Certain movements, such as breathing, ordinarily occur involuntarily, but at times are subject to volition. In presenting this picture, Descartes fails to distinguish adequately between innate reflexes (like breathing) and learned automatic activity (such as walking).

It is somewhat surprising that Descartes says no more than he does, in *Man*, about the conscious instigation of muscular motion. Two things may help explain this omission. First, his primary purpose is to show that many responses conventionally believed to require the soul's intervention can actually occur without it. Second, we hear repeatedly throughout this treatise that the author intends, in a continuation, to discuss the rational soul and its interactions with the body; perhaps in the planned extension we might have heard more on the subject of volition. The related problem of freedom was a central concern of Descartes's, but he was parsimonious in treating its physiological correlates. For his statements on volition, see in Gilson, *ISC*, the listings under *Liberum*, 346, and *Voluntas*, 354. See also Gilson's *La liberté chez Descartes . . .*, Paris, Alcan, 1913.

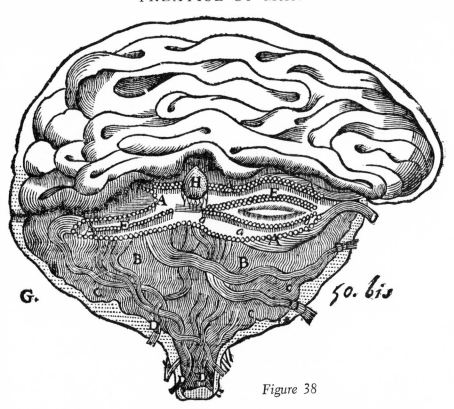

Figure 38

As to dreams, these depend in part on the unequal force that the spirits may have when leaving gland *H* and in part on the impressions that memory entails. Whence dreams differ in only one way from the ideas sometimes formed, as already mentioned, in the imagination of daydreamers. The images formed during sleep can be more distinct and more lively than those that are formed during waking. The reason for this is that when the surrounding parts of the brain are loosened and lax (as seen in diagram 50 [Fig. 39]), a given force can open the tubules (such as 2, 4, and 6) and the pores (such as *a*, *b*, and *c*, which serve to form the images in question) more widely than when these same parts are tense (as seen in the earlier figures). The same reason shows also that if the action of some object impinging on the senses should succeed in reaching the brain during sleep it will not form the same idea that it would during waking, but will form some other more noticeable and perceptible idea: as sometimes if stung by a fly when we sleep, we

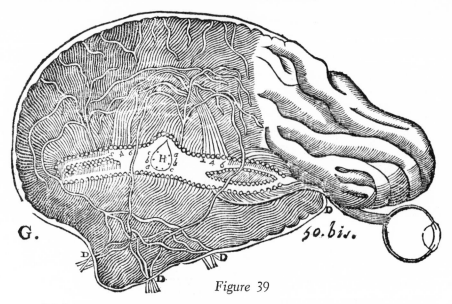

Figure 39

think that someone has stabbed us with a sword; or if we are not adequately covered, we imagine ourselves quite naked; or if we are covered a little too much, we think ourselves weighed down by a mountain [on dreams, see note 131]. [Cl–103; AT–198]

Moreover, during sleep the substance of the brain, being in repose, has leisure to be nourished and repaired, being moistened by the blood contained in the little veins or arteries that are apparent on its external surface. Whence, after some time, the pores having narrowed, the spirits need be less strong to keep the brain substance quite tense (just as the wind need be less strong to inflate a ship's sails when damp than when dry). And yet these spirits [in the brain ventricles] are stronger [during sleep than at other times], inasmuch as the blood producing them is purified while passing and repassing several times through the heart, as already noted above. Whence it follows that this machine must naturally wake itself up after it has slept for some time, just as, reciprocally, it must also go to sleep again after it has been for some time awake; for, during waking, the substance of its brain is dried out and its pores are gradually enlarged by the continual action of the spirits. [Whence it also follows] that, if it [the machine] happens to eat (which hunger will incite it to do at times, if something to eat can be found), the juice of the food, on admixture with blood, renders the latter more coarse and thus makes it produce fewer spirits.[156] [Cl–104; AT–199]

I shall not pause to tell you how the machine is prevented from sleeping by noise, pain, and other actions which, through the mediation of the sense organs, very forcefully move the internal parts of its brain, nor by joy and anger and the other passions that greatly agitate the spirits, nor by dryness of the air which renders the blood more subtle, nor by other similar circumstances. Nor shall I tell you how, on the contrary, silence, sadness, humidity of the

156. The brain exists in two states: waking, in which its fibers are tense, its spirits strong but rapidly exhausted; and sleeping, in which its fibers are lax, its spirits gradually replenished. Exhaustion of strong or forceful spirits in a waking person induces the brain to enter the sleeping state, that is, the state in which its lax fibers can be kept from collapsing together even without an abundant supply of strong spirits. But as this supply increases in the sleeping person, he awakens, that is, he enters the state in which there is force enough to maintain the fibers in a state of greater tension. The reason for the gradual exhaustion of strong spirits during waking is that the spirits' source, the blood, is used rapidly and is not adequately recirculated through the heart for purification and fortification. Also, during waking the admixture of blood with chyle has a coarsening effect and renders the blood less able to produce strong spirits (see note 120).

Before Descartes, many others had speculated about the physical correlates of sleep. His reasoning here is not entirely outside the tradition of Aristotle, for example, who said that "sleep . . . arises from the evaporative loss that is part of the process of nutrition" (for more detail see De somno et vigilia, 456b18–30), and of Galen who said, among other things on the subject, that "sleep befalls us when the native heat is converted into food either through weariness or excessive dryness, or when it cannot leave the body on account of immoderate humidity," De causis pulsuum, K9, 140. Nearer to Descartes in time, Juan Vives ascribed sleep to a vapor derived from food that ascends to the brain, is condensed there, and flows down like a cloud through the whole body, preventing the nerves from functioning as they should (De anima et vita, Basel, 1538, 107). Similarly, Piccolhomini believed that when "sleep comes over us, benign vapors are borne upward from the food and poured into the origins of the nerves preventing the flow of animal spirits into the organs of sensation and spontaneous motion" (Praelectiones, 273). Paré thought of sleep as "repose of the whole body, and chiefly of the animal faculty." He was not sure whether this was due to "a useful moisture imbued into the brain substance which it burdens and weighs down or to a deficiency of spirits which, dissipated by work, render the body no longer able to stand and which constrain the brain to rest so as to replenish these spirits from others derived from food in the stomach." From these and other details. Paré derives some conclusions about the hygiene of sleeping (Oeuvres, 1575, 28–29).

On the whole, the history of ideas about sleep in the century preceding Descartes lacked consistency and concreteness. The examples just mentioned suggest that, without being dominated by any particular theory, Descartes does his reasoning within a tradition that associates sleep in various ways with the physical conditions — especially of the spirits, or fluids — in the brain. For his views elsewhere on sleep, and for Scholastic treatments of it with which he may have been acquainted, see note 131 above and Gilson, ISC, 101 and 271–272.

air, and the like invite sleep. Nor [shall I explain] how too great a loss of blood, or too much fasting or drinking, or such other excesses as have something about them that augments or diminishes the strength of the spirits, can, according to their different temperaments, make the machine either wake too much or sleep too much. Nor how by excessive waking its brain can be weakened and by excessive sleeping grow heavy like that of one who is stupid or out of his senses. Nor [shall I explain] an infinity of other such things: since they all seem easily deducible from things I have already explained. [Cl–105; AT–200]

But before I pass to the description of the rational soul, I still wish you to reflect a little on all that I have just said about this machine; and to consider, firstly, that I have supposed in it only such organs and springs of power as may easily persuade you that wholly similar ones are present both in us and in many nonrational animals as well. For as to those that can clearly be seen, the anatomists have already noted them all. And as to what I have said about the way in which the arteries carry spirits into the head and about the difference between the internal surface of the brain and the middle of its substance, they [the anatomists] will be able to see sufficient visible indications of these so as not to doubt them if they but look a little more closely. No longer will they be able to doubt the little doors or valves that I placed in the nerves where they enter each muscle if they take care [to note] that nature generally has placed such valves wherever in our bodies some matter that enters could tend to escape, as at the entries to the heart, the gall bladder, the throat, and the large intestines, and at the principal divisions of the veins. Concerning the brain, they will be able to imagine nothing more likely than that it is composed of many filaments variously interlaced, seeing that all membranes and fleshy parts seem similarly composed of many fibers or threads [see note 15], and that one observes the same thing in all plants — whence it [a fibrous constitution] seems to be a common property of all bodies that can grow and be nourished by the union and coalition of the particles of other bodies. Finally, as for the rest of the things that I have assumed — things not perceivable by any sense — they are all so simple, so commonplace, and so few, that if you compare them with the diverse composition and marvelous ingenuity that appear in the visible organs, you will suppose rather that I have omitted some that are in us than that I have included some that are not. And knowing that nature always acts by the simplest and easiest means, you will perhaps conclude that the means here proposed are more similar to those she uses than any others that could possibly be found.[157]

I desire you to consider, further, that all the functions that I have attributed to this machine, such as [a] the digestion of food; [b] the beating of the heart and arteries; [c] the nourishment and growth of the members; [d] respiration; [e] waking and sleeping; [f] the reception by the external sense organs of light, sounds, smells, tastes, heat, and all other such qualities; [g] the imprinting of the ideas of these qualities in the organ of common sense and imagination; [h] the retention or imprint of these ideas in the memory; [i] the internal movements of the appetites and passions; and finally [j], the external movements of all the members that so properly follow both the actions of objects presented to the senses and the passions and impressions which are entailed in the memory — I desire you to consider, I say, that these functions imitate those of a real man as perfectly as possible and that they follow naturally in this machine entirely from the disposition of the organs — no more nor less than do the movements of a clock or other automaton, from the arrangement of its counterweights and wheels. Wherefore it is not necessary, on their account, to conceive of any vegetative or sensitive soul or any other principle of movement and life than its blood and its spirits, agitated by the heat of the fire which burns continually in its heart and which is of no other nature than all those fires that occur in inanimate bodies.[158] [Cl–107; AT–202]

157. In the opening paragraph of *Man*, Descartes announces his intention to discuss the body first, then the soul, and then their interactions. *Man* contains only the first section of this tripartite plan. Throughout the work we read repeatedly such statements as "When, as I shall later acknowledge, a soul is bound to this body . . .," but this part of his objective was never realized.

Plato, in discussing the creation of both the cosmos (which he considered to be alive) and the living microcosm (man), depicted the Architect as building the body first (note 64). But this was not to be taken literally; rather, it was an explanatory device to show how the organism "is made." In Descartes's hands, the device is useful especially in emphasizing exactly what *res extensa* (the body) can do independently of *res cogitans* (the soul). See also, on this point, Rothschuh (D., 135, note 1), who believes, on the basis of a statement in the *Discourse* (AT 6:59), that the section on the soul was written but was later lost or destroyed.

158. The central contribution of Descartes to the development of physiology and medicine was the position he adopted with respect to man's soul. Plato's theory, we remember, envisioned two souls or parts of soul: rational and irrational, the latter

having two parts, one passionate and the other appetitive-nutritional. Aristotle recognized four *faculties* of the single soul: nutritive (in all forms of life), sensitive (in all animals and in men), motive (in mobile animals and in man), and rational (in man alone). These ideas in various adaptations still dominated physiological thought at the time of Descartes. As suggested in our introductory material and in note 2, the key to Descartes's whole endeavor was his elimination of all soul faculties except the rational. See, on this point, his *Description of the Body,* AT 11:224–225, and two letters to Regius (May 1641) in the second of which (AT 3:69–372) we read:

> And so, the first thing I disapprove of [in your text] is that you say that man's *soul is triple.* To say this is heresy in my religion. Further, religion notwithstanding, it is contralogical to conceive of "soul" as a genus whose species are "mind," and the "vegetative force" and the "motive force of animals." By *"sensitive soul"* you should understand nothing other than motive force, lest you confuse it with the rational. Moreover this *"motive force"* differs in no special way from "vegetative force"; both differ entirely in kind from the mind. But . . . let me explain it thus.
>
> *Soul* is single in man, and withal rational; nor should one think that any actions are human unless they depend on reason. The power of vegetation and of motion of the body which in plants and animals are called the *vegetative* and *sensitive soul* are also present in man, but ought not be called *souls* in him, since they are not the first principle of his actions and differ entirely in kind from the rational soul.
>
> Moreover *vegetative* force in man [as in animals] is nothing other than a certain constitution of the parts of the body.

Descartes's influence in physiology has been contested and, as we have noted, misunderstood. During the seventeenth century, physiology assumed a new significance as a science by being assimilated into what Dijksterhuis has called the "Mechanization of the World Picture," that is, of cosmology in general. What is often forgotten is that two traditions developed simultaneously — and partly separately — in seventeenth-century physiological mechanics: a metrical-experimental tradition (whose chief early exponent was Harvey) and a rational-philosophical tradition (with Descartes as its most articulate and best-recognized early spokesman).

We have also suggested that both of these traditions were mathematical, but in different senses of the term — the former in its use of numerical measure, the latter in its determination to derive its conclusions from a few self-evident axioms. It is worth noting that neither Harvey nor Borelli entirely relinquished the idea of the soul as a motive cause of vital action. Nor was Descartes's opposition to that idea, though often endorsed, immediately or generally accepted. For a part of this story see T. S. Hall, "Descartes' physiological method," *Journal of the History of Biology* 3 (1970), 53–79. The important point is that ultimately it prevailed. There could be no physiology, as we understand it, based on animistic theories of antiquity or on the new animisms that developed in the eighteenth century under the influence of such thinkers as Stahl, Whytt, and Sauvages de la Croix. For information on their theories see T. S. Hall, *Ideas of Life and Matter,* Chicago, University of Chicago Press, 1969. The soulless physiology practiced by some in the days of Descartes, and explicitly championed by him, was thus delayed, but eventually it was decisive.

Descartes's most articulate statement on the dispensability of the soul as *causa vitae* is contained in the opening sections of his *Description of the Body* (AT 11:223–226, translated by T. S. Hall from the original edition):

> 1. There could be no more fruitful occupation than the effort to understand oneself. The hoped-for use of such understanding looks not only to morals, as many initially suppose, but also and especially to medicine. I believe that, had we studied the latter science thoroughly enough to understand the nature of

our body, we should have found many sure precepts for curing and preventing diseases and, indeed, for slowing the course of old age. And I believe that [had we made such a study] we should not have attributed functions to the soul that in fact depend on the body and only on the arrangement of its organs.

2. But because it has been our experience since childhood that some of our body's movements are obedient to the will, which is a power of the soul, we have tended to believe that the soul was the cause of movements in general. To which belief the ignorance of Anatomy and Mechanics has largely contributed. For, looking only at the outside of the body, we could not imagine that there were organs enough, or force [ressors] enough, within to move it in the many ways in which we see that it moves. And this error was confirmed by our suppositions [a] that dead bodies have the same organs as living from which they differ in nothing but the absence of the soul and [b] that, in dead bodies, motion is stilled.

3. But, if we try to understand our nature more clearly we can see that, inasmuch as our soul is a substance distinct from our body, it is known to us by the single fact that it thinks, that is, by the fact that it understands, wishes, imagines, remembers, and feels; for all these functions are varieties of thought. But other functions sometimes attributed to the soul entail no thought (such as movements of the heart and arteries, digestion of food by the stomach, and the like) and are merely movements of bodies. And, since bodies are ordinarily not moved by the soul but rather by other bodies in motion, such functions are more correctly attributed to the latter than the former.

4. We can also see that when some parts of our body are injured, for example when a nerve is pricked, this makes them no longer as obedient to our will as they customarily have been. Often it even makes them have convulsive movements, movements contrary to the will. This shows that the soul cannot excite any movement in the body unless all the corporeal organs required for this movement are properly arranged. Quite the contrary; when the body has all its organs arranged for a particular movement, there is no need of the soul to produce the movement in question. Consequently, all the movements that we do not experience as depending on thought, should be attributed not to the soul but only to the arrangement of the organs. Indeed, even the movements that we call voluntary proceed principally from the said arrangement of the organs, since they cannot be excited without that arrangement, whatever will toward them we have, and even though it be the soul that determines [which organs are to act].

5. And although all these movements cease in the body when it is dead and when the soul departs, one cannot from that fact infer that it is the soul that produces the movements. One can only infer that the same single cause [a] makes the body no longer proper for producing the movements and [b] makes the soul absent itself from the body.

Admittedly, it is hard to believe that the mere arrangement of the organs is sufficient to produce in us all the movements that are not determined by our thoughts. That is why I shall try to prove it here, and to explain the whole machine of our body in such a way that we shall have no more occasion to think that our soul excites the movements — those which we do not experience to be presided over by our will — than we have to judge that there is a soul in a clock which causes it to show the hours.

It is worth noting that in the above work of his final years, Descartes has dropped the pretense of describing a merely hypothetical analogue of man and admits to describing man himself.

L'Homme
de
René Descartes

L HOMME
DE RENE'
DESCARTES·

PREMIERE PARTIE.

De la Machine de son Corps.

E s hommes seront composez comme nous, d'vne Ame & d'vn Corps; Et il faut que ie vous décriue premierement le corps à part, puis apres l'ame aussi à part : Et enfin que ie vous monstre comment ces deux Natures doiuent estre iointes & vnies, pour composer des hommes qui nous ressemblent.

I.
De quelles parties doit estre composé l'hómme qu'il décrit.

Ie suppose que le Corps n'est autre chose qu'vne statuë ou machine de Terre, que Dieu forme tout exprés, pour la rendre la plus semblable à nous qu'il est possible : En sorte que non seulement il luy donne au dehors la couleur & la figure de tous nos membres, mais aussi qu'il met au dedans toutes les pieces qui sont re-

II.
Que son Corps est vne machine entierement semblable aux nostres.

quifes pour faire qu'elle marche, qu'elle mange, qu'elle refpire, & enfin qu'elle imite toutes celles de nos fon-ctions qui peuuent eftre imaginées proceder de la ma-tiere, & ne dependre que de la difpofition des organes.

Nous voyons des horloges, des fontaines artificielles, des moulins, & autres femblables machines, qui n'é-tant faites que par des hommes ne laiffent pas d'auoir la force de fe mouuoir d'elles-mefmes en plufieurs di-uerfes façons; Et il me femble que ie ne fçaurois ima-giner tant de fortes de mouuemens en celle-cy, que ie fuppofe eftre faite des mains de Dieu, ny luy attribuer tant d'artifice, que vous n'ayez fujet de penfer qu'il y en peut auoir encore dauantage.

Or ie ne m'arrefteray pas à vous décrire les os, les nerfs, les mufcles, les venes, les arteres, l'eftomac, le foye, la rate, le cœur, le cerueau, ny toutes les autres diuerfes pieces dont elle doit eftre compofée; car ie les fuppofe du tout femblables aux parties de noftre Corps qui ont les mefmes noms, & que vous pouuez vous faire monftrer par quelque fçauant Anatomifte, au moins celles qui font affez groffes pour eftre veües, fi vous ne les connoiffez defia affez fuffifamment de vous mefme: Et pour celles qui à caufe de leur petiteffe font inuifi-bles, ie vous les pourray plus facilement & plus claire-ment faire connoiftre, en vous parlant des mouuemens qui en dependent; Si bien qu'il eft feulement icy befoin que i'explique par ordre ces mouuemens, & que ie vous die par mefme moyen qu'elles font celles de nos fon-ctions qu'ils reprefentent.

III.
Comment
　　Premierement les viandes fe digerent dans l'eftomac

de cette machine, par la force de certaines liqueurs, les viandes se digerent dans son estomac. qui se glissant entre leurs parties, les separent, les agitent, & les échauffent, ainsi que l'eau commune fait celles de la chaux viue, ou l'eau forte celles des metaux. Outre que ces liqueurs estant apportées du cœur fort promptement par les arteres, ainsi que ie vous diray cy-apres, ne peuuent manquer d'estre fort chaudes. Et mesme les viandes sont telles pour l'ordinaire qu'elles se pouroient corrompre & échauffer toutes seules, ainsi que fait le foin nouueau dans la grange, quand on l'y serre auant qu'il soit sec.

Et sçachez que l'agitation que reçoiuent les petites parties de ces viandes en s'échauffant, iointe à celle de l'estomac & des boyaux qui les contiennent, & à la disposition des petits filets dont ces boyaux sont composez, fait qu'à mesure qu'elles se digerent, elles descendent peu à peu vers le conduit par où les plus grossieres d'entr'elles doiuent sortir ; & que cependant les plus subtiles & les plus agitées rencontrent çà & là vne infinité de petits trous, par où elles s'écoulent dans les rameaux d'vne grande vene qui les porte vers le foye, & en d'autres qui les portent ailleurs, sans qu'il y ait rien que la petitesse de ces trous qui les separent des plus grossieres ; ainsi que quand on agite de la farine dans vn sas, toute la plus pure s'écoule, & il n'y a rien que la petitesse des trous par où elle passe qui empesche que le son ne la suiue.

Ces plus subtiles parties des viandes estant inégales, IV. Comment le chyle se conuertit en sang. & encore imparfaitement meslées ensemble, composent vne liqueur qui demeureroit toute trouble & toute

blanchaſtre, n'eſtoit qu'vne partie ſe méle incontinent
auec la maſſe du ſang, qui eſt contenuë dans tous les rameaux de la vene nommée Porte (qui reçoit cette liqueur des inteſtins) dans tous ceux de la vene nommée
Caue (qui la conduit vers le cœur) & dans le foye, ainſi
que dans vn ſeul vaiſſeau.

Meſmes il eſt icy à remarquer que les pores du foye
ſont tellement diſpoſez, que lors que cette liqueur entre dedans, elle s'y ſubtiliſe, s'y elabore, y prend ſa couleur, & y acquiert la forme du ſang; tout ainſi que le
ſuc des raiſins noirs, qui eſt blanc, ſe conuertit en vin
clairet, lors qu'on le laiſſe cuuer ſur la raſpe.

V.
Comment
le ſang ſe
chauffe &
ſe dilate dâs
le cœur. Or ce ſang ainſi contenu dans les venes n'a qu'vn ſeul
paſſage manifeſte par où il en puiſſe ſortir, ſçauoir celuy qui le conduit dans la concauité droite du cœur. Et
ſçachez que la chair du cœur contient dans ſes pores vn
de ces feux ſans lumiere, dont ie vous ay parlé cy-deſſus,
qui la rend ſi chaude & ſi ardente, qu'à meſure qu'il entre du ſang dans quelqu'vne des deux chambres ou concauitez qui ſont en elle, il s'y enfle promptement, & s'y
dilate, ainſi que vous pourez experimenter que fera le
ſang ou le laict de quelque animal que ce puiſſe eſtre, ſi
vous le verſez goutte à goutte dans vn vaſe qui ſoit fort
chaud. Et le feu qui eſt dans le cœur de la machine que
ie vous décris, n'y ſert à autre choſe qu'à dilater, é-
chauffer, & ſubtiliſer ainſi le ſang, qui tombe continuellement goutte à goutte, par vn tuyau de la vene
caue, dans la concauité de ſon coſté droit, d'où il s'exhale dans le poulmon; & de la vene du poulmon, que
les Anatomiſtes ont nommé l'*Artere Veneuſe*, dans ſon

autre concauité, d'où il se distribuë par tout le corps.

La chair du poulmon est si rare & si molle, & tousiours VI.
Quel est l'vsage de la respiration en cette machine. tellement rafraischie par l'air de la respiration, qu'à mesure que les vapeurs du sang, qui sortent de la conca- uité droite du cœur, entrent dedans par l'artere que les Anatomistes ont nommé *la vene arterieuse*, elles s'y é- paississent & conuertissent en sang derechef; puis de là tombent goutte à goutte dans la concauité gauche du cœur; où si elles entroient sans estre ainsi derechef é- paissies, elles ne seroient pas suffisantes pour seruir de nourriture au feu qui y est.

Et ainsi vous voyez que la respiration, qui sert seule- ment en cette machine à y épaissir ces vapeurs, n'est pas moins necessaire à l'entretenement de ce feu, que l'est celle qui est en nous, à la conseruation de nostre vie, au moins en ceux de nous qui sont hommes formez: car pour les enfans, qui estans encore au ventre de leurs meres ne peuuent attirer aucun air frais en respirant, ils ont deux conduits qui supléent à ce defaut, l'vn par où le sang de la vene caue passe dans la vene nommée ar- tere, & l'autre par où les vapeurs, ou le sang rarefié de l'artere nommée vene, s'exhalent & vont dans la gran- de artere. Et pour les animaux qui n'ont point du tout de poulmon, ils n'ont qu'vne seule concauité dans le cœur, ou bien s'ils y en ont plusieurs elles sont toutes consecutiues l'vne à l'autre.

Le pouls, ou battement des arteres, depend des onze VII.
Comment se fait le pouls. petites peaux, qui, comme autant de petites portes, fer- ment & ouurent les entrées des quatre vaisseaux qui re- gardent dans les deux concauitez du cœur; car au mo-

ment qu'vn de ces battemens cesse, & qu'vn autre est
prest de commencer, celles de ces petites portes qui
sont aux entrées des deux arteres, se trouuent exactement fermées, & celles qui sont aux entrées des deux
venes se trouuent ouuertes; si bien qu'il ne peut manquer de tomber aussi-tost deux gouttes de sang par ces
deux venes, vne dans chaque concauité du cœur. Puis
ces gouttes de sang se rarefiant, & s'étendant tout d'vn
coup dans vn espace plus grand sans comparaison que
celuy qu'elles occupoient auparauant, poussent & ferment ces petites portes qui sont aux entrées des deux
venes, empeschant par ce moyen qu'il ne descende dauantage de sang dans le cœur, & poussent & ouurent
celles des deux arteres, par où elles entrent promptement & auec effort, faisant ainsi enfler le cœur & toutes les arteres du corps en mesme temps. Mais incontinent apres, ce sang rarefié se condense derechef, ou pe-
netre dans les autres parties; & ainsi le cœur & les arteres se desenflent, les petites portes qui sont aux deux entrées des arteres se referment, & celles qui sont aux entrées des deux venes se rouurent, & donnent passage à
deux autres gouttes de sang, qui font derechef enfler le
cœur & les arteres, tout de mesme que les precedentes.

VIII.
Que c'est
le sang des
arteres qui
sert à la nu-
trition.
　　　　Sçachant ainsi la cause du pouls, il est aysé à entendre
que ce n'est pas tant le sang contenu dans les venes de
cette machine, & qui vient nouuellement de son foye,
comme celuy qui est dans ses arteres, & qui a desia esté
distillé dans son cœur, qui se peut attacher à ses autres
parties, & seruir à reparer ce que leur agitation continuelle, & les diuerses actions des autres corps qui les

enuironnent, en détachent & font fortir ; car le fang
qui eft dans fes venes s'écoule toufiours peu à peu de
leurs extremitez vers le cœur, (& la difpofition de certaines petites portes, ou valvules, que les Anatomiftes
ont remarquées en plufieurs endroits le long de nos
venes, vous doit affez perfuader qu'il arriue en nous
tout le femblable;) mais au contraire celuy qui eft dans
fes arteres eft pouffé hors du cœur auec effort, & à diuerfes petites fecouffes, vers leurs extremitez ; en forte
qu'il peut facilement s'aller ioindre & vnir à tous fes
membres ; & ainfi les entretenir, ou mefme les faire
croiftre, fi elle reprefente le corps d'vn homme qui y
foit difpofé.

Car au moment que les arteres s'enflent, les petites
parties du fang qu'elles contiennent vont choquer çà
& là les racines de certains petits filets, qui, fortans des
extremitez des petites branches de ces arteres, compofent les os, les chairs, les peaux, les nerfs, le cerueau, &
tout le refte des membres folides, felon les diuerfes façons qu'ils fe ioignent ou s'entrelacent: & ainfi elles ont
la force de les pouffer quelque peu deuant foy, & de fe
mettre en leur place : Puis au moment que les arteres fe
defenflent chacune de fes parties s'arrefte où elle fe
trouue, & par cela feul y eft iointe & vnie à celles qu'elle
touche, fuiuant ce qui a efté dit cy-deffus.

I X.
Comment
fe fait la
nutrition
en cette
machine ;
& commēt
elle croift.

Or fi c'eft le corps d'vn enfant que noftre machine
reprefente, fa matiere fera fi tendre, & fes pores fi ay-
féz à élargir, que les parties du fang qui entreront ainfi
en la compofition des membres folides, feront communement vn peu plus groffes que celles en la place de

qui elles se mettront, ou mesme il arriuera que deux où trois succederont ensemble à vne seule, ce qui sera cause de sa croissance. Mais cependant la matiere de ses membres se durcira peu à peu, en sorte qu'apres quelques années ses pores ne se pourront plus tant élargir; & ainsi cessant de croistre elle representera le corps d'vn homme plus aagé.

X.
Que le sang y circule perpetuellement.

Au reste il n'y a que fort peu de parties du sang, qui se puissent vnir à chaque fois aux membres solides en la façon que ie viens d'expliquer; mais la plus-part retournent dans les venes par les extremitez des arteres, qui se trouuent en plusieurs endroits iointes à celles des venes. Et des venes il en passe peut-estre aussi quelques parties en la nourriture de quelques membres; mais la plus-part retournent dans le cœur, puis de là vont derechef dans les arteres; en sorte que le mouuement du sang dans le corps n'est qu'vne circulation perpetuelle.

XI.
Qu'en circulant ainsi, il se separe & se crible.

De plus il y a quelques-vnes des parties du sang qui se vont rendre dans la rate, & d'autres dans la vesicule du fiel; & tant de la rate & du fiel, comme immediatement des arteres, il y en a qui retournent dans l'estomac & dans les boyaux, où elles seruent comme d'eau forte pour ayder à la digestion des viandes; & pource qu'elles y sont apportées du cœur quasi en vn moment par les arteres, elles ne manquent iamais d'estre fort chaudes; ce qui fait que leurs vapeurs peuuent monter facilement par le gosier vers la bouche, & y composer la saliue. Il y en a aussi qui s'écoulent en vrine au trauers de la chair des rognons, ou en sueur & autres excremens au trauers de toute la peau. Et en tous ces lieux,

c'est

c'eſt ſeulement ou la ſituation, ou la figure, ou la peti-
teſſe des pores par où elles paſſent, qui fait que les vnes
y paſſent plutoſt que les autres, & que le reſte du ſang
ne les peut ſuiure; ainſi que vous pouuez auoir veu diuers cribles, qui eſtant diuer-
ſement percez ſeruent à ſeparer diuers grains les vns des
autres.

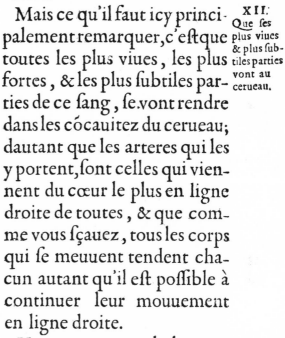

Mais ce qu'il faut icy principalement remarquer, c'eſt que
toutes les plus viues, les plus
fortes, & les plus ſubtiles parties de ce ſang, ſe vont rendre
dans les côcauitez du cerueau;
dautant que les arteres qui les
y portent, ſont celles qui viennent du cœur le plus en ligne
droite de toutes, & que comme vous ſçauez, tous les corps
qui ſe meuuent tendent chacun autant qu'il eſt poſſible à
continuer leur mouuement
en ligne droite.

XII.
Que ſes
plus viues
& plus ſub-
tiles parties
vont au
cerueau.

Voyez par exemple le cœur
A, & penſez que lors que le
ſang en ſort auec effort par
l'ouuerture B, il n'y a aucune
de ſes parties qui ne tende
vers C, où ſont les concauitez
du cerueau; mais que le paſ-

B

fage n'eftant pas affez grand pour les y porter toutes, les plus foibles en font détournées par les plus fortes, qui par ce moyen s'y vont rendre feules.

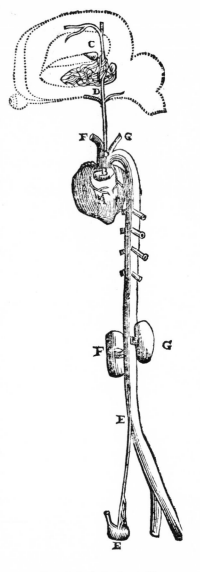

XIII.
Que celles qui n'y peuuent aller vôt aux parties deſtinées à la generatiõ.

Vous pouuez auffi remarquer en paffant, qu'aprés celles qui entrent dans le cerueau, il n'y en a point de plus fortes ny de plus viues, que celles qui fe vont rendre aux vaiffeaux deftinez à la generation. Car par exemple, fi celles qui ont la force de paruenir iufques à D, ne peuuent aller plus auant vers C, à caufe qu'il n'y a pas affez de place pour toutes, elles retournent plutoft vers E, que vers F ny vers G, dautant que le paffage y eft plus droit. En fuite de quoy ie pourrois peut-eftre vous faire voir, comment de l'humeur qui s'affemble vers E, il fe peut former vne autre machine, toute femblable à celle-cy, mais ie ne veux pas entrer plus auant en cette matiere.

XIV.
Des Efprits animaux; ce que

Pour ce qui eft des parties du fang qui penetrent iufqu'au cerueau, elles n'y feruent pas feulement à nourir

& entretenir ſa ſubſtance, mais principalement auſſi à y produire vn certain vent tres ſubtil, ou plutoſt vne flame tres viue & tres pure, qu'on nomme *les Eſprits animaux*. Car il faut ſçauoir que les arteres qui les ap-portent du cœur, apres s'eſtre diuiſées en vne infinité de petites branches, & auoir compoſé ces petits tiſſus, qui ſont eſtendus comme des tapiſſeries au fond des concauitez du cerueau, ſe raſſemblent autour d'vne certaine petite *glande*, ſituée enuiron le milieu de la ſub-ſtance de ce cerueau, tout à l'entrée de ſes concauitez, & ont en cet endroit vn grand nombre de petits trous, par où les plus ſubtiles parties du ſang qu'elles contien-nent, ſe peuuent écouler dans cette glande; mais qui ſont ſi étroits, qu'ils ne donnent aucun paſſage aux plus groſſieres.

Il faut auſſi ſçauoir que ces arteres ne s'arreſtent pas là, mais que s'y eſtant aſſemblées pluſieurs en vne, elles montent tout droit, & ſe vont rendre dans ce grand vaiſſeau qui eſt comme vn Euripe, dont toute la ſuper-ficie exterieure de ce cerueau eſt arroſée. Et de plus il faut remarquer, que les plus groſſes parties du ſang peu-uent perdre beaucoup de leur agitation, dans les de-tours des petits tiſſus par où elles paſſent; dautant qu'el-les ont la force de pouſſer les plus petites qui ſont parmy elles, & ainſi de la leur transferer; mais que ces plus pe-tites ne peuuent pas en meſme façon perdre la leur, dau-tant qu'elle eſt meſme augmentée par celle que leur transferent les plus groſſes, & qu'il n'y a point d'autres corps autour d'elles, auſquels elles puiſſent ſi aiſement la transferer.

D'où il eſt facile à conceuoir que lors que les plus
groſſes montent tout droit vers la ſuperficie exterieure
du cerueau, où elles ſeruent de nourriture à ſa ſubſtance, elles ſont cauſe que les plus petites & les plus agitées
ſe détournent, & entrent toutes en cette glande; qui
doit eſtre imaginée comme vne ſource fort abondante,
d'où elles coulent en meſme temps de tous coſtez dans
les concauitez du cerueau; & ainſi ſans autre preparation, ny changement, ſinon qu'elles ſont ſeparées des
plus groſſieres, & qu'elles retiennent encore l'extreme
viteſſe que la chaleur du cœur leur a donnée, elles ceſ-
ſent d'auoir la forme du ſang, & ſe nomment les eſprits
animaux.

SECONDE PARTIE.

Comment ſe meut la machine de ſon Corps.

XV.
Que les Eſ-
prits ani-
maux ſont
le grand
reſſort qui
fait mou-
uoir cette
machine.

OR à meſure que ces Eſprits entrent ainſi dans les
concauitez du cerueau, ils paſſent de là dans les
pores de ſa ſubſtance, & de ces pores dans les nerfs; où
ſelon qu'ils entrent, ou meſme ſeulement qu'ils tendent
à entrer plus ou moins dans les vns que dans les autres,
ils ont la force de changer la figure des muſcles en qui
ces nerfs ſont inſerez, & par ce moyen de faire mouuoir tous les membres: Ainſi que vous pouuez auoir veu
dans les grottes & les fontaines qui ſont aux jardins de
nos Roys, que la ſeule force dont l'eau ſe meut en ſortant de ſa ſource, eſt ſuffiſante pour y mouuoir diuerſes

machines, & mefme pour les y faire ioüer de quelques
inftrumens, ou prononcer quelques paroles, felon la
diuerfe difpofition des tuyaux qui la conduifent.

Et veritablement l'on peut fort bien comparer les
nerfs de la machine que ie vous décrits, aux tuyaux des
machines de ces fontaines; fes mufcles & fes tendons
aux autres diuers engins & refforts qui feruent à les
mouuoir ; fes efprits animaux a l'eau qui les remuë,
dont le cœur eft la fource, & les concauitez du cerueau font les regars : De plus la refpiration, & autres
telles actions qui luy font naturelles & ordinaires, &
qui dependent du cours des Efprits, font comme les
mouuemens d'vne horloge, ou d'vn moulin, que le
cours ordinaire de l'eau peut rendre continus ; Les objets exterieurs qui par leur feule prefence agiffent contre les organes de fes fens, & qui par ce moyen la determinent à fe mouuoir en plufieurs diuerfes façons, felon
que les parties de fon cerueau font difpofées, font comme des Eftrangers, qui entrans dans quelques-vnes des
grottes de ces fontaines, caufent eux-mefmes fans y
penfer les mouuemens qui s'y font en leur prefence:
Car ils n'y peuuent entrer qu'en marchant fur certains
quarreaux tellement difpofez, que par exemple, s'ils
approchent d'vne Diane qui fe baigne, ils la feront ca-
cher dans des rozeaux ; & s'ils paffent plus outre pour la
pourfuiure, ils feront venir vers eux vn Neptune qui les
menacera de fon Trident ; ou s'ils vont de quelqu'autre
cofté, ils en feront fortir vn monftre marin qui leur vo-
mira de l'eau contre la face, ou chofes femblables, felon le caprice des Ingenieurs qui les ont faites ; Et enfin

XVI.
Belle com-
paraifon
prife des
machines
artificiel-
les.

quand *l'ame raisonnable* fera en cette machine , elle y
aura fon fiege principal dans le cerueau , & fera là comme le fontenier, qui doit eftre dans les regars où fe vont
rendre tous les tuyaux de ces machines, quand il veut
exciter, ou empefcher, ou changer en quelque façon
leurs mouuemens.

XVII.
Sommaire
du refte de
ce traitté.
　　Mais afin que ie vous faffe entendre tout cecy diftinctement, ie veux premierement vous parler de la fabri-
que des nerfs & des mufcles, & vous monftrer comment
de cela feul que les efprits qui font dans le cerueau fe
prefentent pour entrer dans quelques nerfs, ils ont la
force de mouuoir au mefme inftant quelque membre ; Puis ayant touché vn mot de la refpiration, & de
tels autres mouuemens fimples & ordinaires , ie diray
comment les objets exterieurs agiffent contre les organes des fens ; Et apres cela i'expliqueray par le me-
nu tout ce qui fe fait dans les concauitez & dans les pores du cerueau ; comment les efprits animaux y prennent leurs cours ; & quelles font celles de nos fonctions
que cette machine peut imiter par leur moyen: Car fi
ie commençois par le cerueau , & que ie ne fiffe que
fuiure par ordre le cours des efprits, ainfi que i'ay fait
celuy du fang, il me femble que mon difcours ne pourroit pas eftre du tout fi clair.

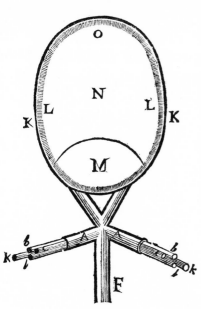

Voyez dóc icy par exemple le nerf A, dont la peau exterieure est comme vn grand tuyau, qui contient plusieurs autres petits tuyaux b, c, k, l, &c. composez d'vne peau interieure plus déliée ; & ces deux peaux sont continuës auec les deux K, L, qui enuelopent le cerueau M, N, o.

Voyez aussi qu'en chacun de ces petits tuyaux, il y a comme vne moëlle, composée de plusieurs filets fort déliez, qui viennent de la propre substance du cerueau N, & dont les extremitez finissent d'vn costé à sa superficie interieure qui regarde ses concauitez, & de l'autre aux peaux & aux chairs contre lesquelles le tuyau qui les contient se termine. Mais pource que cette moëlle ne sert point au mouuement des membres, il me suffit pour maintenant que vous sçachiez qu'elle ne remplit pas tellement les petits tuyaux qui la contiennent, que les esprits animaux n'y trouuent encore assez de place, pour couler facilement du cerueau dans les muscles, où ces petits tuyaux, qui doiuent icy estre comptez pour autant de petits nerfs, se vont rendre.

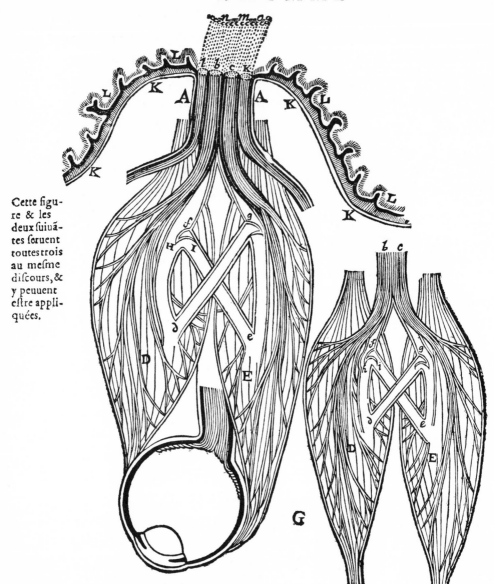

Cette figu-
re & les
deux ſuiuã-
tes ſeruent
toutes trois
au meſme
diſcours, &
y peuuent
eſtre appli-
quées.

Voyez apres cela comment le tuyau, ou petit nerf,
b f, ſe va rendre dans le muſcle D, que ie ſuppoſe eſtre
l'vn de ceux qui meuuent l'œil; & comment y eſtant
il ſe

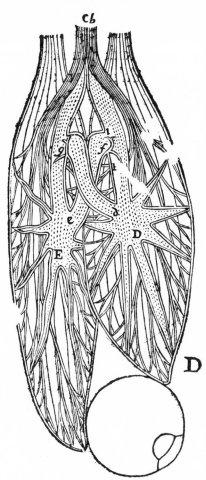

il se diuise en plusieurs bráches, composées d'vne peau lasche, qui se peut étendre, ou élargir & retrecir, selon la quantité des esprits animaux qui y entrent, ou qui en sortent, & dont les rameaux ou les fibres sont tellement disposées, que lors que les esprits animaux entrent dedans, ils font que tout le corps du muscle s'enfle & s'accourcit, & ainsi qu'il tire l'œil auquel il est attaché; comme au contraire lors qu'ils en ressortent ce muscle se desenfle & se rallonge.

De plus, voyez qu'outre le tuyau b f, il y en a encore vn autre, à sçauoir e f, par où les esprits animaux

X X.
Qu'il y a des canaux par où les esprits d'vn musclepeuuent passer dans celuy qui luy est opposé.

peuuent entrer dans le muscle D, & vn autre, à sçauoir d g, par où ils en peuuent sortir. Et que tout de mesme le muscle E, que ie suppose seruir à mouuoir l'œil tout au contraire du precedent, reçoit les esprits animaux du cerueau par le tuyau c g, & du muscle D par par d g, & les renuoye vers D par e f. Et pensez qu'encore qu'il n'y ait aucun passage euident, par où les esprits contenus dans les deux muscles D & E, en puissent

C

fortir, fi ce n'eft pour entrer de l'vn dans l'autre ; toutesfois, pource que leurs parties font fort petites , &
mefme qu'elles fe fubtilifent fans ceffe de plus en plus
par la force de leur agitation , il s'en échappe toufiours
quelques-vnes au trauers des peaux & des chairs de ces
mufcles, mais qu'en reuanche il y en reuient toufiours
auffi quelques autres par les deux tuyaux b f, c g.

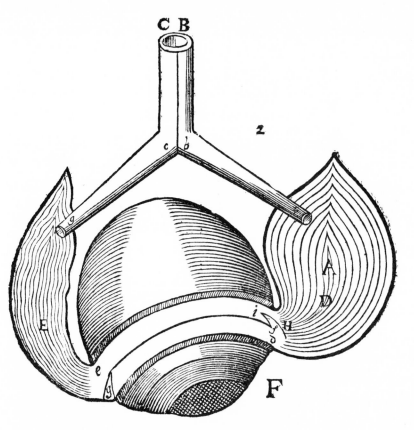

Enfin voyez qu'entre les deux tuyaux b f, e f, il y a
vne certaine petite peau H f i, qui fepare ces deux tuyaux, & qui leur fert comme de porte, laquelle a deux

replis H & i, tellement diſpoſez, que lors que les eſprits animaux qui tendent à deſcendre de b vers H, ont plus de force que ceux qui tendent à monter d'e vers i, ils abbaiſſent & ouurent cette peau, donnans ainſi moyen à ceux qui ſont dans le muſcle E, de couler tres promptement auec eux vers D. Mais lors que ceux qui tendent à monter d'e vers i ſont plus forts, ou ſeulement lors qu'ils ſont auſſi forts que les autres, ils hauſſent & ferment cette peau H f i, & ainſi s'empeſchent euxmeſmes de ſortir hors du muſcle E; au lieu que s'ils n'ont pas de part & d'autre aſſez de force pour la pouſſer, elle demeure naturellement entr'ouuerte. Et enfin que ſi quelquefois les eſprits contenus dans le muſcle D, tendent à en ſortir par d f e, ou d f b, le reply H ſe peut étendre, & leur en boucher le paſſage. Et que tout de meſme entre les deux tuyaux c g, d g, il y a vne petite peau ou valvule g, ſemblable à la precedente, qui demeure naturellement entr'ouuerte, & qui peut eſtre fermée par les eſprits qui viennent du tuyau d g, & ouuerte par ceux qui viennent de c g.

En ſuite de quoy il eſt aiſé à entendre, que ſi les eſprits animaux qui ſont dans le cerueau ne tendent point, ou preſque point, à couler par les tuyaux b f, c g, les deux petites peaux ou valvules f & g demeurent entr'ouuertes, & ainſi que les deux muſcles D & E, ſont laſches & ſans action; dautant que les eſprits animaux qu'ils contiennent, paſſent librement de l'vn dans l'autre, prenans leur cours d'e par f, vers d, & reciproquement de d par g vers e. Mais ſi les eſprits qui ſont dans le cerueau tendent à entrer auec quelque force dans

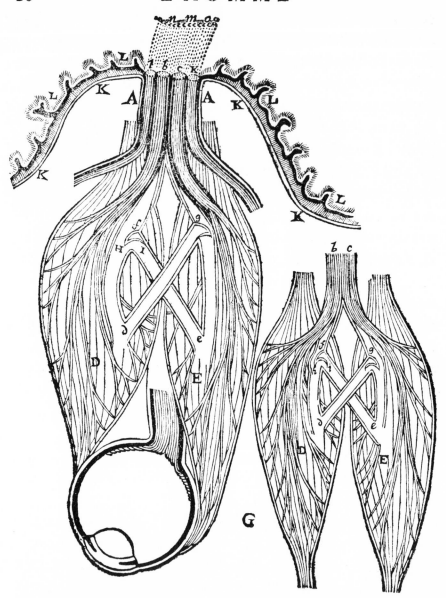

les deux tuyaux b f, c g, & que cette force soit égale
des deux costez, ils ferment aussi-tost les deux passages
g & f, & enflent les deux muscles D & E autant qu'ils

peuuent, leur faifant par ce moyen tenir & arrefter
l'œil ferme en la fituation qu'ils le trouuent.

Puis fi ces efprits qui viennent du cerueau tendent à
couler auec plus de force par b f que par c g, ils ferment la petite peau g, & ouurent f, & ce plus ou moins,
felon qu'ils agiffent plus ou moins fort; au moyen dequoy les efprits contenus dans le mufcle E fe vont rendre dans le mufcle D, par le canal e f; & ce plus ou
moins vifte, felon que la peau f eft plus ou moins ouuerte: Si bien que le mufcle D, d'où ces efprits ne peuuent fortir, s'accourcit, & E fe rallonge; & ainfi l'œil
eft tourné vers D. Comme au contraire, fi les efprits
qui font dans le cerueau tendent à couler auec plus de
force par c g que par b f, ils ferment la petite peau f,
& ouurent g; en forte que les efprits du mufcle D retournent auffi-toft par le canal d g dans le mufcle E,
qui par ce moyen s'accourcit, & retire l'œil de fon
cofté.

Car vous fçauez bien que ces Efprits, eftans comme
vn vent ou vne flame tres fubtile, ne peuuent manquer
de couler tres promptement d'vn mufcle dans l'autre,
fi toft qu'ils y trouuent quelque paffage; encore qu'il
n'y ait aucune autre puiffance qui les y porte, que la feule inclination qu'ils ont à continuer leur mouuement,
fuiuant les loix de la nature. Et vous fçauez outre cela,
qu'encore qu'ils foient fort mobiles & fubtils, ils ne
laiffent pas d'auoir la force d'enfler & de roidir les
mufcles où ils font enfermez; ainfi que l'air qui eft dans
vn balon le durcit, & fait tendre les peaux qui le contiennent.

XXII.
Comment
cette ma-
chine peut
estre meüe
en toutes
les mesmes
façons que
nos corps.

Or il vous eſt aiſé d'appliquer ce que ie viens de dire du nerf A, & des deux muſcles D & E, à tous les autres muſcles & nerfs ; & ainſi d'entendre comment la machine dont ie vous parle, peut eſtre meüe en toutes les meſmes façons que nos corps, par la ſeule force des eſprits animaux qui coulent du cerueau dans les nerfs. Car pour chaque mouuement, & pour ſon contraire, vous pouuez imaginer deux petits nerfs, ou tuyaux, tels que ſont b f, c g, & deux autres tels que ſont d g, e f, & deux petites portes ou valvules telles que ſont н f i, & g.

Et pour les façons dont ces tuyaux ſont inſerez dans les muſcles, encore qu'elles varient en mille ſortes, il n'eſt pas neantmoins mal-aiſé à iuger quelles elles ſont, en ſçachant ce que l'anatomie vous peut apprendre de la figure exterieure, & de l'vſage de chaque muſcle.

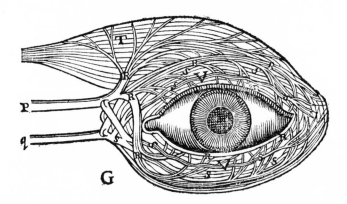

XXIII.
Comment
ſes paupie-
res s'ou-
urent & ſe
ferment.

Car ſçachant par exemple que les paupieres ſont meües par deux muſcles, dont l'vn, à ſçauoir T, ne ſert qu'à ouurir celle de deſſus, & l'autre, à ſçauoir V, ſert alternatiuement à les ouurir & à les fermer toutes deux ; il eſt aifé à penfer qu'ils reçoiuent les efprits par deux tuyaux tels que font p ʀ, & q s; & que l'vn de ces deux tuyaux p ʀ fe va rendre dans ces deux mufcles, & l'autre q s dans l'vn d'eux feulement; Et enfin que les branches ʀ & s eftant quafi inferées en mefme façon dans le mufcle V, y ont toutesfois deux effets tout contraires, à caufe de la diuerfe difpofition de leurs rameaux ou de leurs fibres; ce qui fuffit pour vous faire entendre les autres.

Et mefme il n'eft pas mal-aifé à iuger de cecy, que les efprits animaux peuuent caufer quelques mouue-mens en tous les membres où quelques nerfs fe termi-nent, encore qu'il y en ait plufieurs où les Anatomiftes n'en remarquent aucuns de vifibles ; comme dans la prunelle de l'œil, dans le cœur, dans le foye, dans la veficule du fiel, dans la rate, & autres femblables.

Maintenant pour entendre en particulier comment cette machine refpire, penfez que le mufcle d eft l'vn de ceux qui feruent à hauffer fa poitrine, ou à abbaiffer fon diaphragme, & que le mufcle E eft fon contraire; & que les efprits animaux qui font dans la concauité de fon cerueau marqué m, coulans par le pore ou petit canal marqué n, qui demeure naturellement toufiours ouuert, fe vont rendre d'abord dans le tuyau B F, où abbaiffant la petite peau F, ils font que ceux du mufcle E viennent enfler le mufcle d.

XXIV.
Comment cette ma-chine ref-pire.

Voyez la Figure cy-deffous.

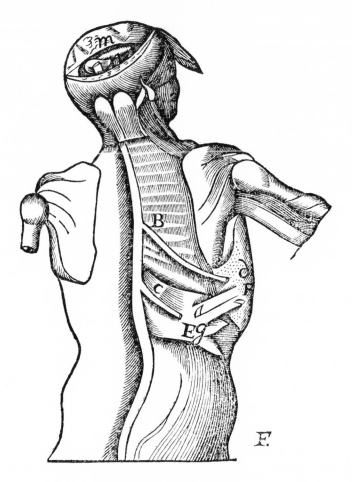

F.

Penſez apres cela qu'il y a certaines peaux autour de
ce muſcle d, qui le preſſent de plus en plus à meſure
qu'il s'enfle, & qui ſont tellement diſpoſées, qu'auant
que tous les eſprits du muſcle E ſoient paſſez vers luy,
elles arreſtent leur cours, & les font comme regorger
par le tuyau B F, en ſorte que ceux du canal n s'en
détournent ; au moyen dequoy s'allans rendre dans le
tuyau c g, qu'ils ouurent en meſme temps, ils font
                                            enfler

enfler le mufcle E, & defenfler le mufcle d; ce qu'ils
continuent de faire auffi long-temps que dure l'impetuofité dont les efprits contenus dans le mufcle d, preffez par les peaux qui l'enuironnent, tendent à en fortir;
Puis, quand cette impetuofité n'a plus de force, ils reprennent d'eux-mefmes leur cours par le tuyau B F, &
ainfi ne ceffent de faire enfler & defenfler alternatiuement ces deux mufcles. Ce que vous deuez iuger auffi
des autres mufcles qui feruent à mefme effet; & penfer
qu'ils font tous tellement difpofez, que quand ce font
les femblables à d qui s'enflent, l'efpace qui contient
les poulmons s'élargit, ce qui eft caufe que l'air entre
dedans, tout de mefme que dans vn foufflet que l'on
ouure; & que quand ce font leurs contraires, cet efpace fe retrecit, ce qui eft caufe que l'air en reffort.

Pour entendre auffi comment cette machine aualle
les viandes qui fe trouuent au fond de fa bouche, penfez que le mufcle d eft l'vn de ceux qui hauffent la ra-
cine de fa langue, & tiennent ouuert le paffage par où
l'air qu'elle refpire doit entrer dans fon poulmon, &
que le mufcle E eft fon contraire, qui fert à fermer ce
paffage, & par mefme moyen à ouurir celuy par où les
viandes qui font dans fa bouche doiuent defcendre dans
fon eftomac, ou bien à hauffer la pointe de fa langue
qui les y pouffe; & que les efprits animaux qui viennent
de la concauité de fon cerueau m, par le pore ou petit
canal n, qui demeure naturellement toufiours ouuert,
fe vont rendre tout droit dans le tuyau B F, au moyen
dequoy ils font enfler le mufcle d: Et enfin que ce mufcle demeure toufiours ainfi enflé, pendant qu'il ne fe

XXV.
Comment
elle aualle
les viandes
qui font
dans fa
bouche.

Seruez
vous de la
figure pre-
cedente, &
que voftre
imagina-
tion fup-
plée à ce
qui man-
que.

D

trouue aucunes viandes au fond de la bouche, qui le
puiſſent preſſer : mais qu'il eſt tellement diſpoſé, que
lors qu'il s'y en trouue quelques-vnes , les eſprits qu'il
contient regorgent auſſi-toſt par le tuyau B F, & font
que ceux qui viennent par le canal n, entrent par le
tuyau c g, dans le muſcle E, où ſe vont auſſi rendre
ceux du muſcle d ; & ainſi la gorge s'ouure, & les viandes deſcendent dans l'eſtomac ; Puis incontinent apres,
les eſprits du canal n reprennent leur cours par B F
comme deuant.

A l'exemple de quoy vous pouuez auſſi entendre comment cette machine peut éternüer , baailler , touſſer,
& faire les mouuemens neceſſaires à rejetter diuers autres excremens.

X X V I.
Comment
elle eſt in-
citée par
les objets
exterieurs à
ſe mouuoir
en pluſieurs
manieres.
Pour entendre apres cela comment elle peut eſtre
incitée, par les objets exterieurs qui frapent les organes de ſes ſens, à mouuoir en mille autres façons tous
ſes membres, penſez que les petits filets, que ie vous ay
deſia tantoſt dit venir du plus interieur de ſon cerueau,
& compoſer la moëlle de ſes nerfs, ſont tellement diſpoſez en toutes celles de ſes parties qui ſeruent d'organe à quelques ſens , qu'ils y peuuent tres facilement
eſtre mûs par les objets de ſes ſens ; & que lors qu'ils y
ſont mûs tant ſoit peu fort, ils tirent au meſme inſtant
les parties du cerueau d'où ils viennent, & ouurent par
meſme moyen les entrées de certains pores , qui ſont
en la ſuperficie interieure de ce cerueau, par où les eſprits animaux qui ſont dans ſes concauitez commencent auſſi-toſt à prendre leur cours, & ſe vont rendre
par eux dans les nerfs, & dans les muſcles, qui ſeruent à

faire en cette machine des mouuemens tout femblables
à ceux aufquels nous fommes naturellement incitez,
lors que nos fens font touchez en mefme forte.

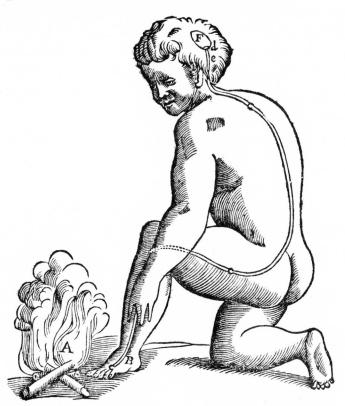

Comme par exemple fi le feu A fe trouue proche du
pié B, les petites parties de ce feu, qui fe meuuent comme vous fçauez tres-promptement, ont la force de
mouuoir auec foy l'endroit de la peau de ce pié qu'elles
touchent; & par ce moyen tirant le petit filet c, c, que
vous voyez y eftre attaché, elles ouurent au mefme inftant l'entrée du pore d, e, contre lequel ce petit filet
fe termine; ainfi que tirant l'vn des bouts d'vne corde,

on fait fonner en mefme temps la cloche qui pend à
l'autre bout.

Or l'entrée du pore ou petit conduit d, e, eſtant ainſi
ouuerte, les eſprits animaux de la concauité F entrent
dedans, & ſont portez par luy, partie dans les muſcles
qui ſeruent à retirer ce pié de ce feu, partie dans ceux
qui ſeruent à tourner les yeux & la teſte pour le regarder, & partie en ceux qui ſeruent à auancer les mains &
à plier tout le corps pour le deffendre.

Mais ils peuuent auſſi eſtre portez par ce meſme conduit d, e, en pluſieurs autres muſcles. Et auant que ie
m'arreſte à vous expliquer plus exactement, en quelle
ſortes les eſprits animaux ſuiuent leur cours par les pores du cerueau, & comment ces pores ſont diſpoſez, ie
veux vous parler icy en particulier de tous les ſens, tels
qu'ils ſe trouuent en cette machine, & vous dire comment ils ſe rapportent aux noſtres.

TROISIESME PARTIE.

*Des ſens exterieurs de cette machine; & comment
ils ſe rapportent aux noſtres.*

XXVII.
De l'attou-
chemen:.

SCachez donc premierement, qu'il y a vn grand
nombre de petits filets ſemblables à c, c, qui commencent tous à ſe ſeparer les vns des autres, dés la ſuperficie interieure de ſon cerueau, d'où ils prennent
leur origine, & qui s'allans de là épandre par tout le
reſte de ſon corps, y ſeruent d'organe pour le ſens de

l'attouchement. Car encore que pour l'ordinaire ce ne foit pas eux qui foient immediatemenr touchez par les objets exterieurs, mais les peaux qui les enuironnent, il n'y a pas toutesfois plus d'apparence de penfer que ce font ces peaux qui font les organes du fens, que de penfer, lors qu'on manie quelque corps eftant ganté, que ce font les gans qui feruent pour le fentir.

Et remarquez qu'encore que les filets dont ie vous parle foient fort déliez, ils ne laiffent pas de paffer feurement depuis le cerueau iufques aux membres qui en font les plus éloignez, fans qu'il fe trouue rien entre deux qui les rompe, ou qui empefche leur action en les preffant, quoy que ces membres fe plient cependant en mille diuerfes façons; dautant qu'ils font enfermez dans les mefmes petits tuyaux qui portent les efprits animaux dans les mufcles, & que ces efprits enflant toufiours quelque peu ces tuyaux, les empefchent d'y eftre preffez; & mefme qu'ils les font toufiours tendre autant qu'ils peuuent, en tirant du cerueau d'où ils viennent, vers les lieux où ils fe terminent.

Or ie vous diray que quand Dieu vnira vne Ame Raifonnable à cette machine, ainfi que ie pretens vous dire cy-apres, il luy donnera fon fiege principal dans le cerueau, & la fera de telle nature, que felon les diuerfes façons que les entrées des pores qui font en la fuperficie interieure de ce cerueau feront ouuertes par l'entremife des nerfs, elle aura diuers fentimens.

XXVIII. De la nature de l'ame, qui doit eftre vnie à cette machine, en ce qui regarde les fens.

Comme premierement, fi les petits filets qui compofent la moëlle de ces nerfs, font tirez auec tant de force, qu'ils fe rompent, & fe feparent de la partie à la-

XXIX. De la douleur, & du chatoüillement.

D iij

quelle ils eſtoient ioints, en ſorte que la ſtructure de toute la machine en ſoit en quelque façon moins accomplie, le mouuement qu'ils cauſeront dans le cerueau donnera occaſion à l'ame, à qui il importe que le lieu de ſa demeure ſe conſerue, d'auoir le ſentiment *de la douleur.*

Et s'ils ſont tirez par vne force preſque auſſi grande que la precedente, ſans que toutesfois ils ſe rompent, ny ſe ſeparent aucunement des parties auſquelles ils ſont attachez, ils cauſeront vn mouuement dans le cerueau, qui rendant témoignage de la bonne conſtitution des autres membres, donnera occaſion à l'ame de ſentir vne certaine volupté corporelle, qu'on nomme *chatoüillement*, & qui, comme vous voyez, eſtant fort proche de la douleur en ſa cauſe, luy eſt toute contraire en ſon effet.

XXX.
Des ſenti-
mens de
rude & de
poly ; de
chaleur &
defroideur,
& autres.

Que ſi pluſieurs de ces petits filets ſont tirez enſemble également, ils feront ſentir à l'ame que la ſuperficie du corps qui touche le membre où ils ſe terminent eſt *polie;* & ils la luy feront ſentir inegale, & qu'elle eſt *rude*, s'ils ſont tirez inegalement.

Que s'ils ne ſont qu'ébranlez quelque peu ſeparement l'vn de l'autre, ainſi qu'ils ſont continuellement par la chaleur que le cœur communique aux autres membres, l'Ame n'en aura aucun ſentiment, non plus que de toutes les autres actions qui ſont ordinaires ; mais ſi ce mouuement eſt augmenté ou diminué en eux par quelque cauſe extraordinaire, ſon augmentation fera auoir à l'Ame le ſentiment *de la chaleur,* & ſa diminution celuy *de la froideur* ; Et enfin ſelon les autres diuerſes façons qu ils feront mûs, ils luy feront fentir toutes les autres qualitez qui appartiennent à l'attouchement en gene-ral, comme *l humidité*, *la féchereffe*, *la pefanteur*, & fem-blables.

Seulement faut-il remarquer qu'encore qu'ils foient fort déliez, & fort aifez à mouuoir, ils ne le font pas toutesfois tellement, qu'ils puiffent rapporter au cer-ueau toutes les plus petites actions qui foient en la na-ture; mais que les moindres qu'ils luy rapportent, font celles des plus groffieres parties des corps terreftres. Et mefme qu'il peut y auoir quelques-vns de ces corps, dont les parties, quoy qu'affez groffes, ne laifferont pas de fe glifler contre ces petits filets fi doucement, qu'el-les les prefferont ou couperont tout à fait, fans que leur action paffe iufqu'au cerueau; Tout de mefme qu'il y a certaines drogues, qui ont la force d'affoupir, ou mef-me de corrompre, ceux de nos membres contre qui elles font appliquées, fans nous en faire auoir aucun fentiment.

XXXI.
De ce qui peut affou-pir le fen-timent.

Mais les petits filets qui compofent la moëlle des nerfs de la langue, & qui feruent d'organe pour le *gouft* en cette machine peuuent eftre mûs par de moindres ac-tions, que ceux qui ne feruent que pour l'attouchement en general, tant à caufe qu'ils font vn peu plus déliez, comme auffi parce que les peaux qui les couurent font plus tendres.

XXXII.
Du gouft, & de fes quatre principales efpeces.

Penfez, par exemple, qu'ils peuuent eftre mûs en quatre diuerfes façons, par les parties des fels, des eaux aigres, des eaux communes, & des eaux de vie, dont ie vous ay cy-deffus expliqué les groffeurs & les figures, &

ainfi qu'ils peuuent faire fentir à l'Ame quatre forte de goufts differens; dautant que les parties des fels eftant feparées l'vne de l'autre, & agitées par l'action de la faliue, entrent de pointe, & fans fe plier, dans les pores qui font en la peau de la langue; celles des eaux aigres s'y coulent de biais, en tranchant ou incifant les plus tendres de fes parties, & obeïffant aux plus groffieres; celles de l'eau douce ne font que fe gliffer par deffus, fans incifer aucunes de fes parties, ny entrer fort auant dans fes pores; & enfin celles de l'eau de vie eftant fort petites y penetrent le plus auant de toutes, & s'y meuuent auec vne tres grande viteffe. D'où il vous eft aifé de iuger comment l'Ame pourra fentir toutes les autres fortes de goufts, fi vous confiderez en combien d'autres façons les petites parties des corps terreftres peuuent agir contre la langue.

XXXIII. Qu'il n'y a que les viandes qui ont du gouft, qui foient propres pour la nourriture.   Mais ce qu'il faut icy principalement remarquer, c'eft que ce font les mefmes petites parties des viandes, qui eftant dans la bouche peuuent entrer dans les pores de la langue, & y émouuoir le fentiment du gouft, lefquelles eftant dans l'eftomac peuuent paffer dans le fang, & de là s'aller ioindre & vnir à tous les membres; & mefme qu'il n'y a que celles qui chatoüillent la langue moderement, & qui pourront par ce moyen faire fentir à l'Ame vn gouft agreable, qui foient entierement propres à cet effet.

Car pour celles qui agiffent trop ou trop peu, comme elles ne fçauroient faire fentir qu'vn gouft trop piquant, ou trop fade, auffi font-elles trop penetrantes, ou trop molles, pour entrer en la compofition du fang, & feruir
à l'en-

à l'entretenement de quelques membres. Et pour celles qui ſont ſi groſſes, ou iointes ſi fort l'vne à l'autre, qu'elles ne peuuent eſtre ſeparées par l'action de la ſaliue, ny aucunement penetrer dans les pores de la langue, pour agir contre les petits filets des nerfs qui y ſeruent pour le gouſt, autrement que contre ceux des autres membres qui ſeruent pour l'attouchement en general, & qui n'ont point auſſi de pores en elles-meſmes, où les petites parties de la langue, ou bien pour le moins celles de la ſaliue dont elle eſt humectée, puiſſent entrer, comme elles ne pourront faire ſentir à l'ame aucun gouſt, ny ſaueur, auſſi ne ſont-elles pas propres pour l'ordinaire à eſtre miſes dans l'eſtomac.

Et cecy eſt ſi generalement vray, que ſouuent à meſure que le temperament de l'eſtomac ſe change, la force du gouſt ſe change auſſi; en ſorte qu'vne viande qui aura coutume de ſembler à l'ame agreable au gouſt, luy pourra meſme quelquefois ſembler fade, ou amere; dont la raiſon eſt que la ſaliue qui vient de l'eſtomac, & qui retient touſiours les qualitez de l'humeur qui y abonde, ſe méle auec les petites parties des viandes qui ſont dans la bouche, & contribuë beaucoup à leur action.

Le ſens de *l'odorat* depend auſſi de pluſieurs petits filets, qui s'auancent de la baze du cerueau vers le nez, au deſſous de ces deux petites parties toutes creuſes, que les Anatomiſtes ont comparées aux bouts des mammelles d'vne femme, & qui ne different en rien des nerfs qui ſeruent à l'attouchement & au gouſt, ſinon qu'ils ne ſortent point hors de la conçauité de la teſte

XXXIV. De l'odorat, & en quoy conſiſtent les bonnes & les mauuaiſes odeurs.

E

qui contient tout le cerueau , & qu'ils peuuent eſtre
mûs par des parties terreſtres encore plus petites que les
nerfs de la langue, tant à cauſe qu'ils ſont vn peu plus
déliez, comme auſſi à cauſe qu'ils ſont plus immediatement touchez par les objets qui les meuuent.

Car vous deuez ſçauoir que lors que cette machine
reſpire, les plus ſubtiles parties de l'air qui luy entrent
par le nez, penetrent par les pores de l'os qu'on nomme ſpongieux , ſinon iuſqu'au dedans des concauitez
du cerueau, pour le moins iuſqu'à l'eſpace qui eſt entre
les deux peaux qui l'enuelopent, d'où elles peuuent reſſortir en meſme temps par le palais ; comme reciproquement quand l'air ſort de la poitrine elles peuuent
entrer dans cet eſpace par le palais, & en reſſortir par
le nez ; & qu'à l'entrée de cet eſpace elles rencontrent
les extremitez de ces petits filets toutes nuës, ou ſeulement couuertes d'vne peau qui eſt extremement déliée, ce qui fait qu'elles n'ont pas beſoin de beaucoup
de force pour les mouuoir.

Vous deuez auſſi ſçauoir que ces pores ſont tellement
diſpoſez, & ſi étroits, qu'ils ne laiſſent paſſer iuſqu'à ces
petits filets , aucunes parties terreſtres qui ſoient plus
groſſes que celles que i'ay cy-deſſus nommées *Odeurs*
pour ce ſujet ; ſi ce n'eſt peut-eſtre auſſi quelques-vnes
de celles qui compoſent les eaux de vie, à cauſe que leur
figure les rend fort penetrantes.

Enfin vous deuez ſçauoir qu'entre ces parties terreſtres
extremement petites, qui ſe trouuent touſiours en plus
grande abondance dans l'air, qu'en aucun des autres
corps compoſez, il n'y a que celles qui ſont vn peu plus

ou moins grofses que les autres, ou qui à raison de leur figure font plus ou moins aifées à mouuoir, qui pourront donner occafion à l'ame d'auoir les diuers fentimens des odeurs: Et mefme il n'y aura que celles en qui ces excez font fort moderez, & temperez l'vn par l'autre, qui luy en feront auoir d'agreables. Car pour celles qui n'agiffent qu'à l'ordinaire, elles ne pourront aucunement eftre fenties; & celles qui agiffent auec trop ou trop peu de force, ne luy pourront eftre que déplaifantes.

Pour les petits filets qui feruent d'organe au fens de l'ouye, ils n'ont pas befoin d'eftre fi déliez que les precedens; mais il fuffit de penfer qu'ils font tellement difpofez au fond des concauitez des oreilles, qu'ils peuuent facilement eftre mûs tous enfemble, & d'vne mefme façon, par les petites fecouffes dont l'air de dehors pouffe vne certaine peau fort déliée, qui eft tenduë à l'entrée de ces concauitez, & qu'ils ne peuuent eftre touchez par aucun autre objet que par l'air qui eft au deffous de cette peau; car ce feront ces petites fecouffes, qui paffans iufqu'au cerueau par l'entremife de ces nerfs, donneront occafion à l'ame de conceuoir l'idée des fons.

XXXV.
De l'ouye; & de ce qui fait le fon.

Et notez qu'vne feule d'entr'elles ne luy pourra faire ouïr autre chofe qu'vn bruit fourd, qui paffe en vn moment, & dans lequel il n'y aura point d'autre varieté, finon qu'il fe trouuera plus ou moins grand, felon que l'oreille fera frappée plus ou moins fort; mais que lors que plufieurs s'entrefuiuront, ainfi qu'on void à l'œil que font les tremblemens des cordes, & des cloches quand elles fonnent, alors ces petites fecouffes compoferont

XXXVI.
En quoy confifte le fon doux ou rude, & tous les tôs de la mufique.

E ij

vn fon, que l'ame iugera plus doux ou plus rude, felon qu'elles feront plus égales ou plus inegales entr'elles ; & qu'elle iugera plus aigu ou plus graue, felon qu'elles feront plus promptes à s'entrefuiure, ou plus tardiues ; en forte que fi elles font de la moitié, ou du tiers, ou du quart, ou d'vne cinquiéme partie &c. plus promptes à s'entrefuiure vne fois que l'autre, elles compoferont vn fon que l'ame iugera plus aigu d'vne octaue, ou d'vne quinte, ou d'vne quarte, ou d'vne tierce majeure &c. Et enfin plufieurs fons mélez enfemble feront accordans ou difcordans, felon qu'il y aura plus ou moins de raport, & qu'il fe trouuera des interualles plus égaux ou plus inegaux, entre les petites fecouffes qui les compofent.

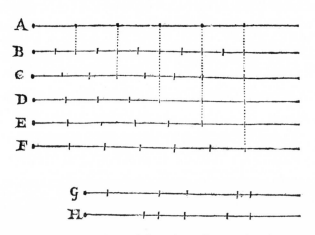

Comme par exemple, fi les diuifions des lignes, A B C, D, E, F, G, H, reprefentent les petites fecouffes qui compofent autant de diuers fons, il eft aifé à iuger que ceux qui font reprefentez par les lignes G & H ne doiuent pas eftre fi doux à l'oreille que les autres ; ainfi

que les parties raboteuses d'vne pierre ne le font pas
tant à l'attouchement, que celle d'vn miroir bien poly.
Et il faut penser que B represente vn son plus aigu que
A, d'vne octaue, C d'vne quinte, D vne quarte, E d'v-
ne tierce majeure, & F d'vn ton aussi majeur ; & remarquer qu'A & B joints ensemble, ou A B C, ou A B D,
ou mesme A B C E font beaucoup plus accordans que
ne font A & F, ou A C D, ou A D E, &c. Ce qui me
semble suffire pour monstrer comment l'ame, qui sera
en la machine que ie vous décris, pourra se plaire à vne
musique qui suiura toutes les mesmes regles que la nô-
tre ; & comment mesme elle pourra la rendre beaucoup
plus parfaite ; au moins si l'on considere, que ce ne font
pas absolument les choses les plus douces, qui font les
plus agreables aux sens, mais celles qui les chatoüillent
d'vne façon mieux temperée ; ainsi que le sel & le vi-
naigre font souuent plus agreables à la langue que l'eau
douce ; Et c'est ce qui fait que la musique reçoit les
tierces & les sextes, & mesme quelquefois les dissonances, aussi bien que les vnissons, les octaues, & les quintes.

Il reste encore le sens *de la veüe*, que i'ay besoin d'expliquer vn peu plus exactement que les autres, à cause
qu'il sert dauantage à mon suiet. Ce sens depend aussi
en cette machine de deux nerfs, qui doiuent sans doute
estre composez de plusieurs petits filets, les plus délicz,
& les plus aisez à mouuoir qui puissent estre ; dautant
qu'ils font destinez à rapporter au ceruau ces diuerses
actions des parties du second element, qui, suiuant ce
qui a esté dit cy-dessus, donneront occasion à l'ame,
quand elle sera vnie à cette machine, de conceuoir les

XXXVII.
De la veue.

L iij

diuerſes idées des couleurs & de la lumiere.

XXXVIII
De la ſtru-
ꝃure de
l'œil ; & en
quoy elle
ſert à la vi-
ſion.
Mais pource que la ſtructure de l'œil aide auſſi à cet effet, il eſt icy beſoin que ie la décriue ; & pour plus grande facilité ie taſcheray de le faire en peu de mots, en laiſſant tout à deſſein pluſieurs particularitez ſuper‑ fluës, que la curioſité des Anatomiſtes y remarque.

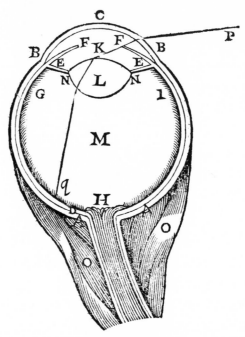

A B C eſt vne peau aſſez dure & épaiſſe, qui compoſe comme vn vaze rond, dans lequel toutes les autres par‑ ties de l'œil ſont contenuës. D E F en eſt vne autre plus déliée, qui eſt tendüe ainſi qu'vne tapiſſerie au dedans de la precedente. G H I eſt le nerf, dont les petits filets H G, H I, eſtant épars tout autour depuis H iuſques à G & I, couurent entierement le fond de l'œil. K, L, M, ſont trois ſortes de glaires, ou humeurs, extremement

claires & tranſparentes qui rempliſſent tout l'eſpace
contenu au dedans de ces peaux, & qui ont chacune la
figure que vous voyez icy repreſentée.

En la premiere peau la partie B C B eſt tranſparente,
& vn peu plus voûtée que le reſte ; & la refraction des
rayons qui entrent dedans s'y fait vers la perpendicu-
laire ; En la deuxiéme peau¹, la ſuperficie interieure de
la partie E F, qui regarde le fond de l'œil, eſt toute
noire & obſcure, & elle a au milieu vn petit trou rond,
qui eſt ce qu'on nomme la *prunelle*, & qui paroiſt ſi noir
au milieu de l'œil, quand on le regarde par dehors. Ce
trou n'eſt pas touſiours de meſme grandeur, car la partie E F de la peau dans laquelle il eſt, nageant librement dans l'humeur K, qui eſt fort liquide, ſemble
eſtre comme vn petit muſcle, qui s'élargit ou s'étrecit
par la direction du cerueau, ſelon que l'vſage le re-
quiert.

La figure de l'humeur marqué L qu'on nomme *l'humeur cryſtalline*, eſt ſemblable à celle de ces verres, que
i'ay décrits au traitté de la Dioptrique, par le moyen
deſquels tous les rayons qui viennent d'vn certain point
ſe raſſemblent à vn autre certain point ; & ſa matiere
eſt moins molle, ou plus ferme, & cauſe par conſequent
vne plus grande refraction, que celle des deux autres
humeurs qui l'enuironnent.

E, N, ſont de petits filets noirs, qui viennent du dedans de la peau D, E, F, & qui embraſſent tout autour
cette humeur cryſtalline ; qui ſont comme autant de
petits tendons, par le moyen deſquels ſa figure ſe peut
changer, & ſe rendre vn peu plus platte, ou plus voûtée,

felon qu'il eft de befoin. Enfin o, o, font fix ou fept
mufcles attachez à l'œil par dehors, & qui le peuuent
mouuoir tres facilement &tres promptement de tous
coftez.

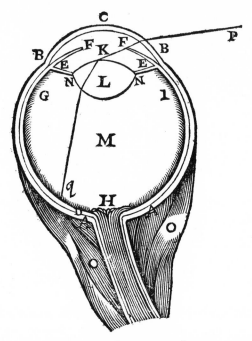

XXXIX.
Ce que fait
la tranfpa-
rence des
trois hu-
meurs.

Or la peau B C B, & les trois humeurs K, L, M,
eftant fort claires & tranfparentes, n'empefchent point
que les rayons de la lumiere, qui entrent par le trou de
la prunelle, ne penetrent iufqu'au fond de l'œil, où eft
le nerf, & qu'ils n'agiffent auffi facilement contre luy,
comme s'il eftoit tout à fait à découuert ; & elles feruent
à le preferuer des iniures de l'air, & des autres corps exterieurs, qui le pourroient facilement offenfer s'ils le
touchoient ; & de plus à faire qu'il demeure fi tendre &
fi delicat, que ce n'eft pas merueille qu'il puiffe eftre
meu

meu par des actions si peu sensibles, comme sont celles
que ie prens icy pour *les couleurs.*

La courbure qui est en la partie de la premiere peau,
marquée B C B, & la refraction qui s'y fait, est cause
que les rayons qui viennent des objets qui sont vers les
costez de l'œil, peuuent entrer par la prunelle; & ainsi
que sans que l'œil se remüe, l'Ame pourra voir plus grád
nombre d'objets, qu'elle ne pourroit faire sans cela:
car par exemple si le rayon P B K q ne se courboit pas
au point B, il ne pourroit passer entre les points F, F,
pour paruenir iusques au nerf.

X L.
Ce que fait
la courbure
de la pre-
miere peau.

La refraction qui se fait en l'humeur crystalline sert
à rendre la vision plus forte, & ensemble plus distincte:
Car vous deuez sçauoir, que la figure de cette humeur
est tellement compassée, eu égard aux refractions qui
se font dans les autres parties de l'œil, & à la distance
des objets, que lors que la veüe est dressée vers quelque
point determiné d'vn objet, elle fait que tous les rayons
qui viennent de ce point, & qui entrent dans l'œil par le
trou de la prunelle, se rassemblent en vn autre point au
fond de l'œil, iustement contre l'vne des parties du nerf
qui y est, & empesche par mesme moyen, qu'aucuns
des autres rayons qui entrent dans l'œil, ne touche la
mesme partie de ce nerf.

X L I.
La refrac-
tió de l'hu-
meur cry-
stalline réd
la vision
plus forte,
& plus di-
stincte.

Par exemple l'œil estant disposé à regarder le point R, la disposition de l'humeur crystalline fait
que tous les rayons R N S, R L S
&c. s'assemblent iustement au
point S, & empesche par mesme
moyen , qu'aucun de ceux qui
viennent des points T & X &c.
n'y paruiennent ; car elle assemble aussi tous ceux du point T enuiron le point V, ceux du point X
enuiron le point Y, & ainsi des
autres ; au lieu que s'il ne se fai-
soit aucune refraction dans cet
œil, l'objet R n'enuoyeroit qu'vn
seul de ses rayons au point S, & les
autres s'épandroient çà & là en
tout l'espace V, Y ; & de mesme
les points T & X, & tous ceux qui
sont entre deux , enuoyeroient

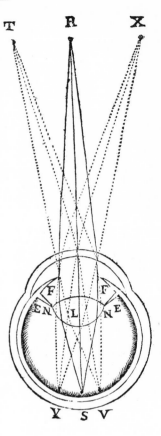

chacun vn de leurs rayons vers ce mesme point S.

Or il est bien euident que l'objet R doit agir plus
fort contre la partie du nerf qui est à ce point S, lors
qu'il y enuoye grand nombre de rayons, que s'il n'y en
enuoyoit qu'vn seul ; & que cette partie du nerf S, doit
rapporter plus distinctement & plus fidelement au cerueau l'action de cet objet R, lors qu'elle ne reçoit des
rayons que de luy seul, que si elle en receuoit de diuers
autres.

XLII.　　La couleur noire tant de la superficie interieure de la

peau E F, que des petits filets E N, ſert auſſi à rendre
la viſion plus diſtincte : car ſuiuant ce qui a eſté dit cy-
deſſus de la nature de cette couleur, elle amortit la force des rayons qui ſe reflechiſſent du fond de l'œil vers
le deuant, & empeſche que de là ils ne retournent derechef vers le fond de l'œil, où ils pourroient apporter de
la confuſion. Par exemple les rayons de l'objet X don-
nant au point Y contre le nerf qui eſt blanc, ſe reflechiſſent de là de tous coſtez vers N & vers F, d'où ils

La couleur
noire qui
eſt au de-
dãs de l'œil
ſert auſſi à
rendre la
viſion plus
diſtincte.

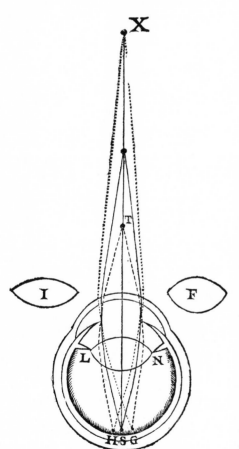

pourroient derechef ſe
reflechir vers S & vers
V, & y troubler l'action
des points R & T, ſi les
corps N & F n'eſtoient
pas noirs.

Le changement de fi-
gure qui ſe fait en l'hu-
meur cryſtalline, ſert à
ce que les objets qui ſõt
à diuerſes diſtances puiſſent peindre diſtincte-
ment leurs images au
fond de l'œil : car ſuiuãt
ce qui a eſté dit au traitté de la Dioptrique, ſi
par exemple l'humeur
L N eſt de telle figure,
qu'elle faſſe que tous les
rayons qui partent du
point R aillent iuſtemét

XLIII.
Le change-
ment de fi-
gure de
l'humeur
cryſtalline
ſert auſſi à
la diſtin-
ction des
images.

G ij

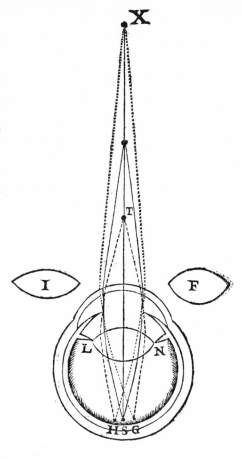

toucher le nerf au point
S , la mefme hûmeur,
fans eftre changée , ne
pourra faire que ceux du
point T , qui eft plus
proche, ou du point X,
qui eft plus éloigné , y
aillent auffi ; mais elle
fera que le rayon T L
ira vers H, & T N vers
G ; & au contraire que
X L ira vers G , & X N
vers H, & ainfi des au-
tres. Si bien que pour reprefenter diftinctement
le point X , il eft befoin
que toute la figure de
cette humeur N L fe
change , & qu'elle de-
uienne vn peu plus plat-
te , comme celle qui eft
marquée I ; Et pour reprefenter le point T, il eft befoin qu'elle deuienne vn peu plus voûtée , comme celle
qui eft marquée F.

X LIV.
Le change-
ment de
grandeur
en la pru-
nelle , fert à
moderer la
force de la
vifion.   Le changement de grandeur qui arriue à la prunelle
fert à moderer la force de la vifion ; car il eft befoin
qu'elle foit plus petite quand la lumiere eft trop viue,
afin qu'il n'entre pas tant de rayons dans l'œil que le
nerf en puiffe eftre offenfé ; & qu'elle foit plus grande
quand la lumiere eft trop foible, afin qu'il y en entre

affez pour eftre fentis. Et de plus, pofant que la lumiere demeure égale, il eft befoin que la prunelle foit plus
grande, quand l'objet que l'œil regarde eft éloigné,
que quand il eft proche : car par exemple, s'il n'entre
qu'autant de rayons du point R, par la prunelle de
l œil 7, qu'il en faut pour pouuoir eftre fentis; il eft
befoin qu'il en entre tout autant dans l'œil 8, & par
confequent que fa prunelle foit
plus grande.

La petiteffe de la prunelle fert
auffi à rendre la vifion plus diftincte; car vous deuez fçauoir
que quelque figure que puiffe a-
uoir l'humeur cryftalline, il eft
impoffible qu'elle faffe que les
rayons qui viennent de diuers
points de l'objet s'affemblent
tous exactement en autant d'autres diuers points : Mais que fi
ceux du point R, par exemple,
s'affemblent iuftement au point
S, il n'y aura du point T, que
ceux qui paffent par la circonfe-
rence & par le centre de l'vn des
cercles qu'on peut décrire fur la
fuperficie de cette humeur cryftalline, qui fe puiffent affembler
exactement au point V; & par
confequent que les autres, qui
feront dautant moindres en nom-

X L V.
Que la pe-
titeffe de la
prunelle
fert auffi à
rendre la
vifion plus
diftincte.

G iij

bre que la prunelle fera plus petite, allans toucher le nerf en d'autres points, ne pourront manquer d'y apporter de la confufion ; d'où vient que fi la vifion d'vn mefme œil eft moins forte vne fois que l'autre, elle fera auffi moins diftincte, foit que cela vienne de l'éloignement de l'objet, foit de la debilité de la lumiere ; parce que la prunelle eftant plus grande quand elle eft moins forte, cela rend auffi la vifion plus confufe.

X L VI.
Que l'ame
ne pourra
voir diftin-
ctement
qu'vn feul
point de
l'objet.

Voyez la
figure pa-
ge 42.

De là vient auffi que l'ame ne pourra iamais voir tres diftinctement qu'vn feul point de l'objet à chaque fois, fçauoir, celuy vers lequel toutes les parties de l'œil feront dreffées pour lors, & que les autres luy paroiftront dautant plus confus, qu'ils feront plus éloignez de celuy-cy : Car, par exemple, fi les rayons du point R s'affemblent tous exactement au point S, ceux du point X s'affembleront encore moins exactement vers Y, que ceux du point T ne s'affembleront vers V ; & il faut iuger ainfi des autres, à mefure qu'ils font plus éloignez du point R. Mais les mufcles o, o, (cy deuant reprefentez dans la premiere figure de l'œil page 40.) tournant l'œil tres promptement de tous coftez, feruent à fuppléer à ce defaut : car ils peuuent en moins de rien l'appliquer fucceffiuement à tous les points de l'objet, & ainfi faire que l'Ame les puiffe voir tous diftinctement l'vn apres l'autre.

X L V I.
Quels ob-
jets font
agreables
ou defa-
greables à
la veüe.

Ie n'adjoute pas icy particulierement ce que c'eft qui pourra donner occafion à cette Ame de conceuoir toutes les differences des couleurs, car i'en ay defia affez parlé cy-deffus ; Et ie ne dis pas auffi quels objets de la veüe luy doiuent eftre agreables ou defagreables ; car

de ce que i'ay expliqué des autres fens, il vous eft facile
à entendre que la lumiere trop forte doit offenfer les
yeux, & que la moderée les doit recréer ; & qu'entre
les couleurs, la verte, qui confifte en l'action la plus
moderée (qu'on peut nommer par analogie la proportion d'vn à deux) eft comme l'octaue entre les confo-
nances de la mufique, ou le pain entre les viandes que
l'on mange, c'eft à dire celle qui eft la plus vniuerfelle-
ment agreable ; Et enfin que toutes ces diuerfes couleurs de la mode, qui recréent fouuent beaucoup plus
que le vert, font comme les accords & les paffages d'vn
air nouueau, touché par quelque excellent ioüeur de
luth, ou les ragoufts d'vn bon cuifinier, qui chatoüillent bien dauantage le fens, & luy font fentir d'abord
plus de plaifir, mais auffi qui le laffent beaucoup plutoft
que ne font les objets fimples & ordinaires.

Seulement faut-il encore que ie vous die ce que c'eft
qui donnera moyen à l'Ame de fentir la fituation, la figure, la diftance, la grandeur, & autres femblables
qualitez qui ne fe rapportent pas à vn feul fens en particulier, ainfi que font celles dont i'ay parlé iufques icy ;
mais qui font communes à l'attouchement & à la veüe,
& mefme en quelque façon aux autres fens.

XLVIII.
Comment
on voit la
fituation,
la figure, la
diftance, &
la grandeur
des objets.

Remarquez donc premierement que fi la main A,
par exemple, touche le corps C, les parties du cerueau
B, d'où viennent les petits filets de fes nerfs, feront autrement difpofées, que fi elle en touchoit vn qui fuft
d'autre figure, ou d'autre grandeur, ou fitué en vne
autre place; & ainfi que l'Ame pourra connoiftre par
leur moyen la fituation de ce corps, & fa figure & fa
grandeur, & toutes les autres femblables qualitez. Et
que tout de mefme, fi l'œil D eft tourné vers l'objet
E, l'Ame pourra connoiftre la fituation de cet objet,
dautant que les nerfs de cet œil feront difpofez en vne
autre

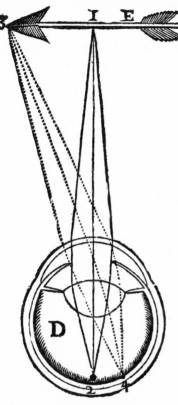

autre forteque s'il eſtoit tourné vers ailleurs ; & qu'elle
pourra connoiſtre ſa figure,
dautant que les rayons du
point 1. s'aſſemblans au point
2, contre le nerf nommé optique, & ceux du point 3. au
point 4, & ainſi des autres, y
en traceront vne qui ſe rap-
portera exactement à la ſien-
ne ; & qu'elle pourra connoi-
tre la diſtance du point 1. par
exemple, dautant que la diſpoſition de l'humeur cryſtalline ſera d'autre figure, pour
faire que tous les rayons qui
viennent de ce point s'aſſemblent au fond de l'œil iuſtement au point 2, que ie ſup-
poſe en eſtre le milieu, que s'il en eſtoit plus proche ou
plus éloigné, ainſi qu'il a tantoſt eſté dit ; Et de plus
qu'elle connoiſtra celle du point 3. & de tous les autres
dont les rayons entreront dans l'œil en meſme temps;
pour ce que l'humeur cryſtalline eſtant ainſi diſpoſée,
les rayons de ce point 3 ne s'aſſembleront pas ſi iuſtement au point 4, que ceux du point 1 au point 2, &
ainſi des autres ; & que leur action ne ſera pas du-tout
ſi forte à proportion, ainſi qu'il a auſſi tantoſt eſté dit.
Et enfin que l'Ame pourra connoiſtre la grandeur des
objets de la veüe, & toutes leurs autres ſemblables qua-

G

litez, par la seule connoissance qu'elle aura de la distance & de la situation de tous leurs points; comme aussi
reciproquement elle iugera quelquefois de leur distance, par l'opinion qu'elle aura de leur grandeur.

Remarqués aussi que
si les deux mains f & g,
tiennent chacune vn
baston, i & h, dont el-
les touchent l'objet K,
encore que l'ame igno-
re d'ailleurs la longueur
de ces bastons; toutes-
fois pource qu'elle sçau-
ra la distance qui est
entre les deux points f
& g, & la grandeur des
angles f g h, & g f i, elle
pourra cónoistre com-
me par vne geometrie
naturelle où est l'objet
K. Et tout de mesme, si
les deux yeux L & M
sont tournez vers l'ob-
jet N, la grandeur de la
ligne L M, & celle des
deux angles L M N,
M L N luy feront con-
noistre où est le point
N.

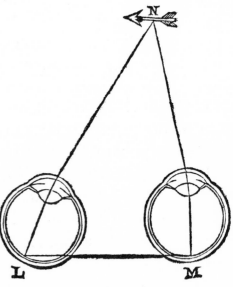

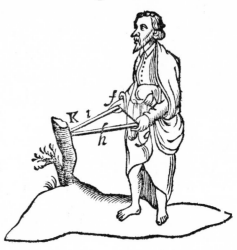

XLIX.   Mais elle pourra aussi

affez fouuent fe tromper en tout cecy ; car premierement fi la fituation de la main, ou de l'œil, ou du doigt
eſt contrainte par quelque cauſe exterieure, elle ne
s'accordera pas ſi exactement auec celle des petites parties du cerueau d'où viennent les nerfs, comme ſi elle
ne dependoit que des muſcles ; Et ainſi l'Ame, qui ne
la ſentira que par l'entremiſe des parties du cerueau, ne
manquera pas pour lors de ſe tromper.

Qu'on s'y
peut ſouuēt
tromper; &
pourquoy
l'on voit
quelque-
fois les ob-
jets dou-
bles.

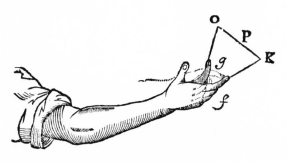

Comme par exemple ſi la main f eſtant de ſoy diſpoſée à ſe tourner vers o, ſe trouue contrainte par quelque force exterieure à demeurer tourné vers K, les parties du cerueau d'où viennent ſes nerfs ne feront pas
tout à fait diſpoſées en meſme ſorte, que ſi c'eſtoit par
la force de ſes muſcles que la main fuſt ainſi tournée
vers K ; ny auſſi en meſme ſorte, que ſi elle eſtoit veritablement tournée vers o ; mais d'vne façon moyenne entre ces deux, ſçauoir en meſme ſorte que ſi elle
eſtoit tournée vers P: Et ainſi la diſpoſition que cette
contrainte donnera aux parties du cerueau fera iuger à
l'Ame que l'objet K eſt au point P, & qu'il eſt autre que
celuy qui eſt touché par la main g.

G ij

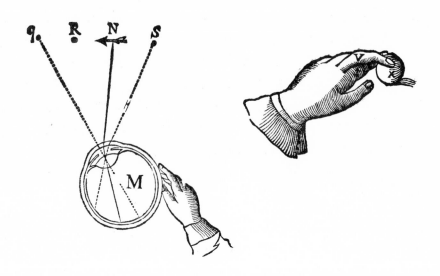

Tout de mesme si l'œil M est détourné par force de l'objet N, & disposé comme s'il deuoit regarder vers q, l'Ame jugera que l'œil est tourné vers R; &pource qu'en cette situation les rayons de l'objet N entreront dans l'œil, tout de mesme que feroient ceux du point S, si l'œil estoit veritablement tourné vers R, elle croira que cet objet N est au point S, & qu'il est autre que celuy qui est regardé par l'autre œil.

Tout de mesme aussi, les deux doigts t & v touchans la petite boule X, feront iuger à l'Ame qu'ils en touchent deux differentes, à cause qu'ils sont croisez & retenus par contrainte hors de leur situation naturelle.

De plus si les rayons, ou autres lignes, par l'entremise desquelles les actions des objets éloignez passent vers les sens, sont courbées, l'Ame, qui les supposera communement estre droites, en tirera occasion de se trom-

I.
Pourquoy ils paroissent autrement situez qu'ils ne sont, &

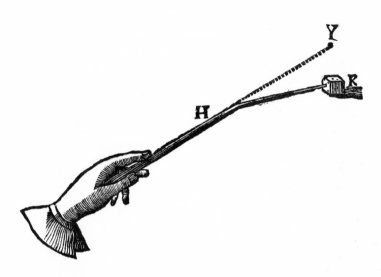

per; Comme par exemple, ſi le baſton H Y eſt courbé
vers K, il ſemblera à l'Ame que l'objet K, que ce
baſton touche, eſt vers Y. Et ſi l'œil L reçoit les rayons
de l'objet N au trauers du verre Z, qui les courbe, il
ſemblera a l'Ame que cet objet eſt vers A. Et tout de
meſme ſi l'œil B reçoit les rayons du point D, au trauers du verre c, que ie ſuppoſe les plier tous en meſme
façon que s'ils venoient du point E, & ceux du point F,
comme s'ils venoient du point G, & ainſi des autres, il
ſemblera à l'Ame que l'objet D F H, eſt auſſi éloigné
& auſſi grand que paroiſt E G I.

Voyez les deux Figures cy-deſſous.

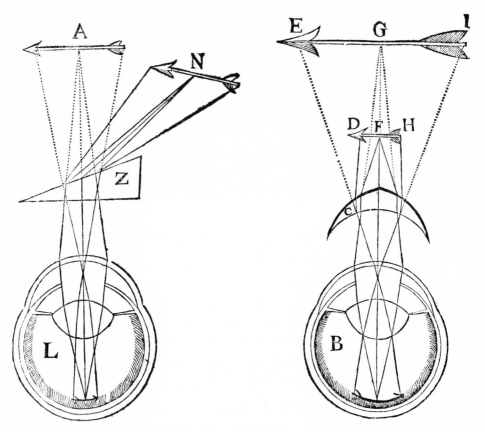

Voyez la
fig. p. 50

LI.
Que tous
les moyens
de connoi-
ſtre la di-
ſtance des
ob ets ſont
incertains.

Et pour concluſion il faut remarquer , que tous les moyens que l'Ame aura pour connoiſtre la diſtance des objets de la veüe, ſont incertains : car pour les angles L M N, M L N, & leurs ſemblables, ils ne changent quaſi plus ſenſiblement, quand l'objet eſt à quinze ou vingt piez de diſtance ; Et pour la diſpoſition de l'humeur cryſtalline, elle change encore moins ſenſiblement, ſi toſt que l'objet eſt plus de trois ou quatre piez loin de l'œil ; Et enfin pour ce qui eſt de iuger des éloignemens, par l'opinion qu'on a de la grandeur des ob-

jets, ou pour ce que les rayons qui viennent de leurs diuers points, ne s'assemblent pas si exactement au fond
de l'œil les vns que les autres, l'exemple des tableaux
de perspectiue nous monstre assez combien il est facile
de s'y tromper : Car lors que leurs figures sont plus petites que nous ne nous imaginons qu'elles doiuent estre,
& que leurs couleurs sont vn peu obscures, & leurs lineamens vn peu confus, cela fait qu'elles nous paroissent de beaucoup plus éloignées & plus grandes qu'elles
ne sont.

Or apres vous auoir ainsi expliqué les cinq sens exterieurs, tels qu'ils sont en cette machine, il faut aussi que
ie vous die quelque chose de certains sentimens interieurs qui s'y trouuent.

QVATRIESME PARTIE.

Des sens interieurs qui se trouuent en cette machine.

Lors que les liqueurs, que i'ay dit cy-dessus seruir
comme d'eau forte dans son estomac, & y entrer
sans cesse de toute la masse du sang par les extremitez
des arteres, n'y trouuent pas assez de viandes à dissoudre
pour occuper toute leur force, elles la tournent contre
l'estomac mesme, & agitant les petits filets de ses nerfs
plus fort que de coutume, font mouuoir les parties du
cerueau d'où ils viennent : Ce qui sera cause que l'Ame
estant vnie à cette machine conceura l'idée generale *de*

LII.
De la faim,
& d'où viét
l'appetit de
manger de
certaines
viandes.

la faim. Et ſi ces liqueurs ſont diſpoſées à employer plutoſt leur action contre certaines viandes particulieres
que contre d'autres, ainſi que l'eau forte commune
diſſout plus aiſement les metaux que la cire, elles agi-
ront auſſi d'vne façon particuliere contre les nerfs de
l'eſtomac, laquelle ſera cauſe que l'Ame conceura pour
lors l'appetit de manger de certaines viádes, plutoſt que
d'autres. (*Hîc notari poteſt mira huius machinæ conformatio,
quod fames oriatur ex ieiunio: ſanguis enim circulatione acrior
ſic, & ita liquor ex eo in ſtomachum veniens neruos magis vellicat; idque modo peculiari, ſi peculiaris ſit conſtitutio ſangui-
nis; vnde pica mulierum.* L'on peut icy remarquer la ſtructure admirable de cette machine, qui eſt telle que la
faim luy vient d'auoir eſté trop long-temps ſans manger; dont la raiſon eſt que le ſang ſe ſubtiliſe & deuient
plus acre par la circulation; d'où il arriue que la liqueur
qui va des arteres dans ſon eſtomac agite & picote plus
fort que de coutume les nerfs qui y ſont, & meſme
qu'elle les agite d'vne certaine façon particuliere, ſi la
conſtitution du ſang ſe trouue auſſi auoir quelque choſe de particulier : Et c'eſt de là que viennent ces appetits deſordonnez, ou ces enuies des femmes groſſes.) Or
ces liqueurs s'aſſemblent principalement au fond de
l'eſtomac, & c'eſt là qu'elles cauſent le ſentiment de la
faim.

LIII.
De la ſoif,
& cóment
elle eſt ex-
citée.
Mais il monte auſſi continuellement pluſieurs de leurs
parties vers le goſier, & lors qu'elles n'y viennent pas en
aſſez grande abondance pour l'humecter, & remplir
ſes pores en forme d'eau, elles y montent ſeulement en
forme d'air, ou de fumée, & agiſſant pour lors contre
ſes

ſes nerfs d'autre façon que de coutume, elles cauſent vn mouuement dans le cerueau, qui donnera occaſion à l'Ame de conceuoir l'idée *de la ſoif.*

Ainſi, lors que le ſang qui va dans le cœur eſt plus pur & plus ſubtil, & s'y embraſe plus facilement qu'à l'ordinaire, il diſpoſe le petit nerf qui y eſt, en la façon qui eſt requiſe pour cauſer le ſentiment *de la ioye;* & en celle qui eſt requiſe pour cauſer le ſentiment *de la triſteſſe,* quand ce ſang a des qualitez toutes contraires.

LIV.
De la ioye & de la triſteſſe, & des autres ſentimens interieurs.

Et de cecy vous pouuez aſſez entendre ce qu'il y a en cette machine qui ſe rapporte à tous les autres ſentimens interieurs qui ſont en nous; ſi bien qu'il eſt temps que ie commence à vous expliquer comment les eſprits animaux ſuiuent leur cours dans les concauitez & dans les pores de ſon cerueau, & quelles ſont les fonctions qui en dependent.

Si vous auez iamais eu la curioſité de voir de prés les orgues de nos Egliſes, vous ſçauez comment les ſouf flets y pouſſent l'air en certains receptacles, qui ce me ſemble ſont nommez à cette occaſion les porte-vents, & comment cet air entre de là dans les tuyaux, tantoſt dans les vns, tantoſt dans les autres, ſelon les diuerſes façons que l'organiſte remüe ſes doigts ſur le clauier. Or vous pouuez icy conceuoir que le cœur & les arteres qui pouſſent les eſprits animaux dans les concauitez du cerueau de noſtre machine, ſont comme les ſouf flets de ces orgues, qui pouſſent l'air dans les porte-vents; & que les objets exterieurs, qui, ſelon les nerfs qu'ils remüent, font que les eſprits contenus dans ces concauitez entrent de là dans quelques-vns de ces po-

LV.
Belle comparaiſon, qui explique d'où dependent toutes les fonctions de cette machine.

H

res, font comme les doigts de l'organifte, qui, felon les touches qu'ils preffent, font que l'air entre des porte-vents dans quelques tuyaux. Et comme l'harmonie des orgues ne dépend point de cet arrangement de leurs tuyaux que l'on voit par dehors, ny de la figure de leurs porte-vents, ou autres parties, mais feulement de trois chofes, fçauoir de l'air qui vient des foufflets, des tuyaux qui rendent le fon, & de la diftribution de cet air dans les tuyaux; Ainfi ie veux vous aduertir, que les fonctions dont il eft icy queftion, ne dependent aucunement de la figure exterieure de toutes ces parties vifibles que les Anatomiftes diftinguent en la fubftance du cerueau, n'y de celle de fes concauitez; mais feulement des ef-prits qui viennent du cœur, des pores du cerueau par où ils paffent, & de la façon que ces efprits fe diftri-buent dans ces pores: Si bien qu'il eft feulement icy befoin que ie vous explique par ordre tout ce qu'il y a de plus confiderable en ces trois chofes.

LVI. Que les diuerfes inclinations naturelles dependent de la diuerfi-té des ef-prits.

Premierement, pour ce qui eft des Efprits Animaux, ils peuuent eftre plus ou moins abondans, & leurs par-ties plus ou moins groffes, & plus ou moins agitées, & plus ou moins égales entr'elles vne fois que l'autre; & c'eft par le moyen de ces quatre differences que toutes les diuerfes humeurs ou inclinations naturelles qui font en nous (au moins entant qu'elles ne dependent point de la conftitution du cerueau, ny des affections parti-culieres de l'Ame) font reprefentées en cette machine. Car fi ces Efprits font plus abondans que de coutume, ils font propres à exciter en elle des mouuemens tout femblables à ceux qui témoignent en nous *de la bonté*, *de la liberalité & de l'amour* ; Et de femblables à ceux qui témoignent en nous *de la confiance ou de la hardieffe* , fi leurs parties font plus fortes & plus groffes ; & *de la conftance* , fi auec cela elles font plus égales en figure , en force , & en groffeur ; & *de la promptitude , de la diligence, & du defir* , fi elles font plus agitées ; & *de la tranquillité d'efprit* , fi elles font plus égales en leur agitation. Comme au contraire ces mefmes Efprits font propres à exciter en elle des mouuemens tout femblables à ceux qui témoignent en nous *de la malignité , de la timidité , de l'inconftance , de la tardiueté , & de l'inquietude* , fi ces mefmes qualitez leur defaillent.

Et fçachez que toutes les autres humeurs ou inclinations naturelles font dependantes de celles-cy; Comme, *l'humeur joycufe* eft compofée de la promptitude & de la tranquillité d'efprit ; & la bonté & la confiance feruent à la rendre plus parfaite. *L'humeur trifte* eft compofée de la tardiueté & de l'inquietude , & peut eftre augmentée par la malignité & la timidité. *L'humeur colerique* eft compofée de la promptitude & de l'inquietude , & la malignité & la confiance la fortifient. Enfin , comme ie viens de dire , la liberalité , la bonté , & l'amour dependent de l'abondance des Efprits , & forment en nous cette humeur qui nous rend complaifans & bienfaifans à tout le monde. La curiofité & les autres defirs dependent de l'agitation de leurs parties ; & ainfi des autres.

Mais parce que ces mefmes humeurs , ou du moins les paffions aufquelles elles difpofent , dependent auffi beaucoup des impreffions qui fe font dans la fubftance du cerueau , vous les pourrez cy-aprés mieux entendre;

H ij

& ie me contenteray icy de vous dire les cauſes d'où viennent les differences des Eſprits.

LVII.
Que le ſuc des viandes rēd le ſang ordinaire-ment plus groſſier.
Le ſuc des viandes qui paſſe de l'eſtomac dans les ve-nes, ſe mélant auec le ſang, luy communique touſiours quelques-vnes de ſes qualitez, & entr'autres il le rend ordinairement plus groſſier, quand il ſe méle tout fraiſ-chement auec luy ; en ſorte que pour lors les petites parties de ce ſang que le cœur enuoye vers le cerueau, pour y compoſer les Eſprits Animaux, ont coutume de n'eſtre pas ſi agitées, ny ſi fortes, ny ſi abondantes ; & par conſequent de ne rendre pas le corps de cette ma-chine ſi leger, ny ſi alaigre, comme il eſt quelque temps apres que la digeſtion eſt acheuée, & que le meſme ſang ayant paſſé & repaſſé pluſieurs fois dans le cœur eſt deuenu plus ſubtil.

LVIII.
Que l'air de la reſpi-ſation rend les eſprits plus vifs & plus agitez.
L'air de la reſpiration ſe mélant auſſi en quelque fa-çon auec le ſang, auant qu'il entre dans la concauité gauche du cœur, fait qu'il s'y embraſe plus fort, & y produit des Eſprits plus vifs & plus agitez en temps ſec qu'en temps humide ; ainſi qu'on experimente que pour lors toute ſorte de flame eſt plus ardente.

LIX.
Que le foye bien diſpo-ſé les rend plus abon-dans & plus également agitez.
Lors que le foye eſt bien diſpoſé, & qu'il elabore par-faitement le ſang qui doit aller dans le cœur, les Eſprits qui ſortent de ce ſang en ſont dautant plus abondans, & plus également agitez ; & s'il arriue que le foye ſoit preſſé par ſes nerfs, les plus ſubtiles parties du ſang qu'il contient, montans incontinent vers le cœur, produi-ront auſſi des Eſprits plus abondans & plus vifs que de coutume, mais non pas ſi également agitez.

LX.
Si le fiel, qui eſt deſtiné à purger le ſang de celles de

ſes parties qui ſont les plus propres de toutes à eſtre embraſées dans le cœur, manque à faire ſon deuoir, ou
qu'eſtant reſſerré par ſon nerf, la matiere qu'il contient regorge dans les venes, les Eſprits en ſeront dautant plus vifs, & auec cela plus inegalement agitez.

Que le fiel
les rēd plus
vifs, & plus
inegalemēt
agitez.

Si la rate, qui au contraire eſt deſtinée à purger le
ſang de celles de ſes parties qui ſont les moins propres
à eſtre embraſées dans le cœur, eſt mal diſpoſée, ou
qu'eſtant preſſée par ſes nerfs, ou par quelqu'autre
corps que ce ſoit, la matiere qu'elle contient regorge
dans les venes, les Eſprits en ſeront dautant moins abondans, & moins agitez, & auec cela plus inegalement
agitez.

LXI.
Que la rate
les rend
moins a-
bondans &
moins agi-
tez.

Enfin tout ce qui peut cauſer quelque changement
dans le ſang en peut auſſi cauſer dans les Eſprits. Mais
par deſſus tout, le petit nerf qui ſe termine dans le cœur,
pouuant dilater & reſſerrer, tant les deux entrées par
où le ſang des venes & l'air du poulmon y deſcend, que
les deux ſorties par où ce ſang s'exhale & s'elance dans
les arteres, peut cauſer mille differences en la nature
des eſprits; ainſi que la chaleur de certaines lampes fermées, dont ſe ſeruent les Alchymiſtes, peut eſtre moderée en pluſieurs façons, ſelon qu'on ouure plus ou
moins, tantoſt le conduit par où l'huile, ou autre aliment de la flame y doit entrer, & tantoſt celuy par où
la fumée en doit ſortir.

LXII.
Que le pe-
tit nerf du
cœur cauſe
le plus de
diuerſité
dans les eſ-
prits.

CINQVIESME PARTIE.

De la ſtruĉture du cerueau de cette machine ; Et comment les eſprits s'y
diſtribuent pour cauſer ſes moumemens & ſes ſentimens.

LXIII.
De la ſtru-
ĉure du
cerueau de
cette ma-
chine.

SEcondement, pour ce qui eſt des pores du cerueau,
ils ne doiuent pas eſtre imaginez autrement que
comme les interualles qui ſe trouuent entre les filets de
quelque tiſſu : car en effet tout le cerueau n'eſt autre
choſe qu'vn tiſſu compoſé d'vne certaine façon particuliere, que ie taſcheray icy de vous expliquer.

Cette figu-
re & la ſui-
uante peu-
uent eſtre
appliquées
au meſme
diſcours.

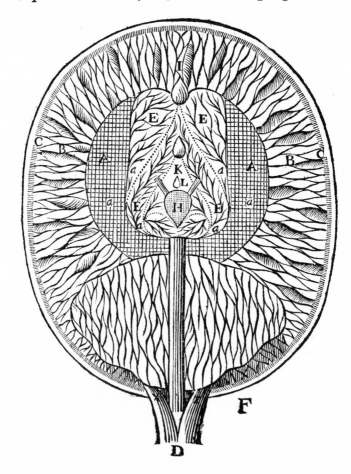

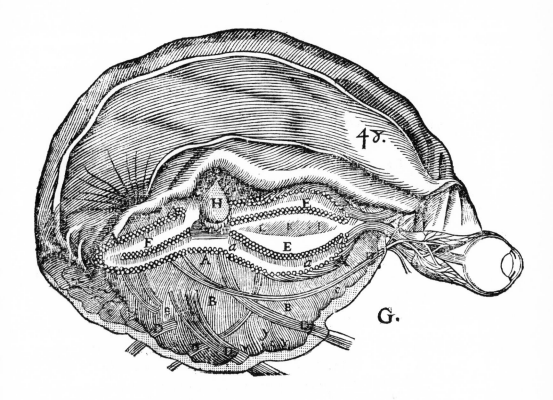

Conceuez fa fuperficie A A, qui regarde les concauitez E E, comme vn rezeüil ou laflis affez épais &
preffé, dont toutes les mailles font autant de petits tuyaux par où les efprits animaux peuuent entrer, & qui re-
gardans toufiours vers la glande H, d'où fortent ces
efprits, fe peuuent facilement tourner çà & là vers les
diuers points de cette glande ; comme vous voyez qu'ils

Cette figu-
re fera nō-
mée cy-
aprés fig.
M.

font tournez autrement
en la 48 qu'en la 49 fi-
gure ; & penſez que de
chaque partie de ce re-
zeüil il ſort pluſieurs filets fort déliez, dont les
vns ſont ordinairement
plus longs que les au-
tres ; Et qu'aprés que ces
filets ſe ſont diuerſemét
entrelacez en tout l'eſ-
pace marqué B , les plus
longs deſcendent vers

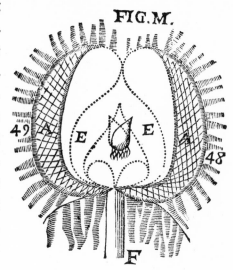

FIG. M.

D , puis de là compoſans la moëlle des nerfs ſe vont é-
pandre par tous les membres.

    Penſez auſſi que les principales qualitez de ces petits
filets ſont de pouuoir aſſez facilement eſtre pliez en toutes ſortes de façons par la ſeule force des eſprits qui les
touchent, &, quaſi comme s'ils eſtoient faits de plomb
ou de cire, de retenir touſiours les derniers plis qu'ils
ont receus, iuſqu'à ce qu'on leur en imprime de contraires.

    Enfin penſez que les pores dont il eſt icy queſtion ne
ſont autre choſe que les interualles que ſe trouuent entre ces filets, & qui peuuent eſtre diuerſement élargis
& retrecis, par la force des eſprits qui entrent dedans,
ſelon qu'elle eſt plus ou moins grande, & qu'ils ſont plus
ou moins abondans ; & que les plus courts de ces filets
ſe vont rendre en l'eſpace c, c, où chacun ſe termine
contre l'extremité de quelqu'vn des petits vaiſſeaux
qui

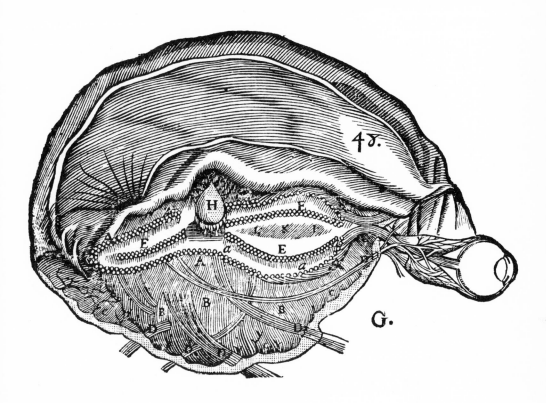

qui y font, & en reçoit fa nourriture.

Troifiémement ; mais afin que ie puiffe plus commo-
dement expliquer toutes les particularitez de ce tiffu, il
faut icy que ie commence à vous parler de la diftribution de ces Efprits.

Iamais ils ne s'arreftent vn feul moment en vne place,
mais à mefure qu'ils entrent dans les concauitez du cerueau E E, par les trous de la petite glande marquee H,

LXIV.
Comment
fe fait la di-
ftribution
des efprits;
& d'où viét
l'eternüe-
ment & l'é-
bloüiffe-
ment ou
vertige.

I

ils tendent d'abord vers ceux des petits tuyaux a, a, qui
leur font le plus directement oppofez; Et fi ces tuyaux
a, a, ne font pas affez ouuerts pour les receuoir tous, ils
reçoiuent au moins les plus fortes & les plus viues de
leurs parties, pendant que les plus foibles & fuperflües
font repouffées vers les conduits I, K, L, qui regardent
les narines, & le palais; à fçauoir les plus agitées vers I,
par où, quand elles ont encore beaucoup de force, &
qu'elles n'y trouuent pas le paffage affez libre, elles fortent quelquefois auec tant de violence, qu'elles chatoüillent les parties interieures du nez, ce qui caufe
l'Eternüment; Puis les autres vers K & vers L, par où
elles peuuent facilement fortir, pource que les paffages
y font fort larges; où fi elles y manquent, eftant con-
traintes de retourner vers les petits tuyaux a, a, qui font
en la fuperficie interieure du cerueau, elles caufent
auffi-toft vn ébloüiffement, ou vertige, qui trouble les
fonctions de l'imagination.

Et notez en paffant que ces plus foibles parties des
Efprits, ne viennent pas tant des arteres qui s'inferent
dans la glande H, comme de celles qui fe diuifans en
mille branches fort déliées tapiffent le fond des concauitez du cerueau. Notez auffi qu'elles fe peuuent aife-
ment épaiffir en pituite, non pas iamais eftant dans le
cerueau, fi ce n'eft par quelque grande maladie, mais
en ces larges efpaces qui font au deffous de fa baze, entre les narines & le gofier; Tout de mefme que la fumée fe conuertit facilement en fuye, dans les tuyaux
des cheminées, mais non pas iamais dans le foyer où
eft le feu.

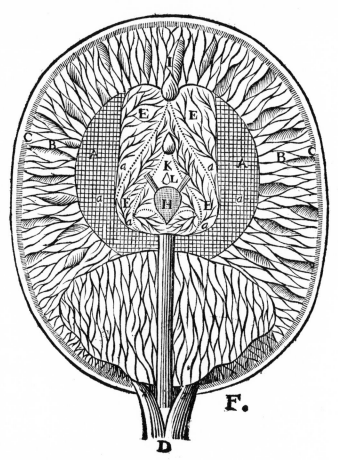

Notez auſſi que lors que ie dis que les Eſprits en ſortant de la glande H, tendent vers les endroits de la ſuperficie interieure du cerueau, qui leur ſont le plus directement oppoſez, ie n'entens pas qu'ils tendent toû-
jours vers ceux qui ſont vis à vis d'eux en ligne droite;
mais ſeulement vers ceux, où la diſpoſition qui eſt pour
lors dans le cerueau les fait tendre.

Or la ſubſtance du cerueau eſtant molle & pliante, LXV.
ſes concauitez ſeroient fort étroites, & preſque toutes Quelle dif-
feren e il y

I ij

a entre le
cerueau
d'vn hôme
qui veille,
& celuy
d'vn hôme
qui dort.

fermées, ainſi qu'elles paroiſſent dans le cerueau d'vn homme mort, s'il n'entroit dedans aucuns Eſprits; mais la ſource qui produit ces Eſprits eſt ordinairement ſi abondante, qu'à meſure qu'ils entrent dans ces conca-uitez, ils ont la force de pouſſer tout autour la matiere qui les enuironne, & de l'enfler, & par ce moyen de faire tendre tous les petits filets des nerfs qui en vien-nent; ainſi que le vent, eſtant vn peu fort, peut enfler les voiles d'vn nauire, & faire tendre toutes les cordes auſquelles ils ſont attachez; D'où vient que pour lors cette machine eſtant diſpoſée à obeïr à toutes les ac-tions des Eſprits, repreſente le corps *d'vn homme qui veille;* Ou du moins ils ont la force d'en pouſſer ainſi &

Cette figu-
re va eſtre
nommée
fig N.

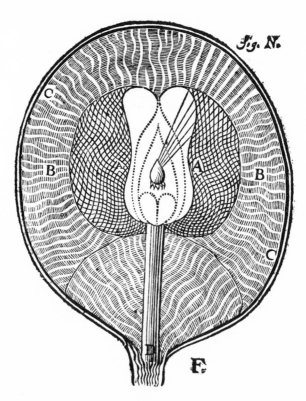

Fig. N.

faire tendre quelques parties , pendant que les autres
demeurent libres & lasches, ainsi que font celles d'vn
voile, quand le vent est vn peu trop foible pour le remplir ; Et pour lors cette machine represente le corps d'vn
homme *qui dort*, & qui a *diuers songes* en dormant. Ima-
ginez-vous par exemple que la difference qui est entre

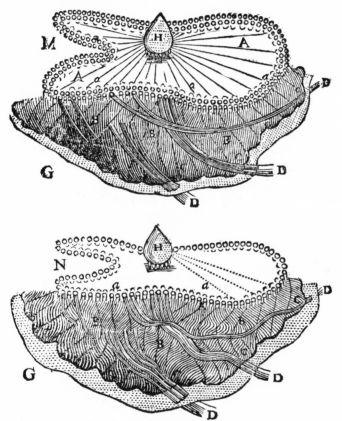

les deux figures M, & N, est la mesme qui est entre le
cerueau d'vn homme qui veille, & celuy d'vn homme
qui dort, & qui réue en dormant.

Ces deux
fig. se voyet
aussi page:
66. & 68.

   Mais auant que ie vous parle plus particulierement du

fommeil & des *fonges*, il faut que ie vous faſſe icy conſi-
derer tout ce qui ſe fait de plus remarquable dans le
cerueau, pendant le temps de la veille, à ſçauoir comment s'y forment les idées des objets, dans le lieu deſtiné pour *l'Imagination*, & pour *le ſens commun*, comment
elles ſe reſeruent dans *la memoire*, & comment elles cauſent le mouuement de tous les membres.

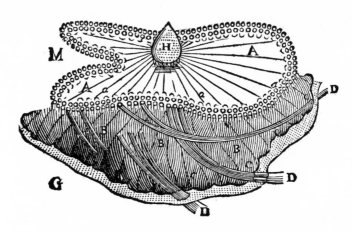

LXVI.
Comment
ſe forment
les idées
des objets
dans le lieu
deſtiné à
l'imagina-
tion, & au
ſens com-
mun.

Vous pouuez voir en la figure marquée M, que les
Eſprits qui ſortent de la glande H, ayant dilaté la partie du cerueau marquée A, & entr'ouuert tous ſes pores,
coulent de là vers B, puis vers C, & enfin vers D, d'où
ils ſe répandent dans tous ſes nerfs, & tiennent par ce
moyen tous les petits filets, dont ces nerfs & le cerueau
ſont compoſez, tellement tendus, que les actions qui
ont tant ſoit peu la force de les mouuoir, ſe communi-
quent facilement de l'vne de leurs extremitez iuſques
à l'autre, ſans que les detours des chemins par où ils paſſent les en empeſchent.

LXVII.   Mais afin que ces détours ne vous empeſchent pas

auſſi de voir clairement, comment cela ſert à former les Que les fi-
gures des
objets ſe
tracēt auſſi
en la ſuper-
ficie inte-
rieure du
cerueau. idées des objets qui frapent les ſens, regardez en la fi-
gure cy-ioínte les petits filets 12, 34, 56, & ſemblables,
qui compoſent le nerf optique, & ſont étendus depuis
le fond de l'œil 1, 3, 5, iuſques à la ſuperficie interieure
du cerueau 2, 4, 6 ; Et penſez que ces filets ſont telle-
ment diſpoſez, que ſi les rayons qui viennent par exem-
ple du point A de l'objet vont preſſer le fond de l'œil,

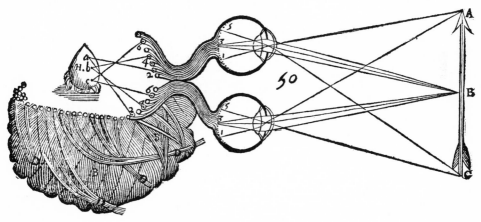

au point 1. ils tirent par ce moyen tout le filet 12, & aug-
mentent l'ouuerture du petit tuyau marqué 2 ; Et tout de
meſme, que les rayons qui viennent du point B, aug-
mentent l'ouuerture du petit tuyau 4, & ainſi des au-
tres. En ſorte que, comme les diuerſes façons dont les
points, 1, 3, 5, ſont preſſez par ces rayons, tracent dans
le fond de l'œil vne figure qui ſe rapporte à celle de l'ob-
jet A B C, ainſi qu'il a eſté dit cy-deſſus, il eſt euident
que les diuerſes façons, dont les petits tuyaux 2, 4, 6,
ſont ouuerts, par les filets 12, 34, 56, &c. la doiuent auſſi

Cette fig.
ſera cy-
apres dite
fig. 50.

tracer en la superficie interieure du cerueau.

LXVIII.
Qu'il s'en
trace aussi
sur la glan-
de, qui se
raportent à
celle des
objets.

Pensez apres cela que les Esprits qui tendent à entrer dans chacun des petits tuyaux 2, 4, 6, & semblables, ne viennent pas indifferemment de tous les points qui sont en la superficie de la glande H, mais seulement de quelqu'vn en particulier; Et que ce sont ceux qui viennent, par exemple du point a, de cette superficie, qui tendent à entrer dans le tuyau 4, & ceux des points b, & c, qui tendent à entrer dans les tuyaux 4, & 6, & ainsi des autres. En sorte qu'au mesme instant que l'ouuerture de ces tuyaux deuient plus grande, les Esprits commencent à sortir plus librement & plus viste qu'ils ne faisoient auparauant, par les endroits de cette glande qui les regardent; Et que comme les diuerses façons dont les tuyaux 2, 4, 6, sont ouuerts, tracent vne figure qui se rapporte à celle de l'objet A, B, C, sur la superficie interieure du cerueau, ainsi celle dont les esprits sortent des points a, b, c, la tracent sur la superficie de cette glande.

LXIX.
Que ces fi-
gures ne
sont que
les diuerses
impressiōs
que reçoi-
uent les es-
prits en
sortant de
la glande.

Et notez que par ces figures, ie n'entens pas seulement icy les choses qui representent en quelque sorte la position des lignes & des superficies des objets, mais aussi toutes celles, qui, suiuant ce que i'ay dit cy-dessus, pourront donner occasion à l'Ame de sentir le mouuement, la grandeur, la distance, les couleurs, les sons, les odeurs, & autres telles qualitez; & mesmes celles qui luy pourront faire sentir le chatoüillement, la douleur, la faim, la soif, la joye, la tristesse, & autres telles passions. Car il est facile à entendre, que le tuyau 2 par exemple sera ouuert autrement par l'action que i'ay dit

causer

cauſer le ſentiment de la couleur rouge , ou celuy du chatoüillement , que par celle que i'ay dit cauſer le ſentiment de la couleur blanche , ou bien celuy de la douleur ; & que les Eſprits qui ſortent du point a , tendront diuerſement vers ce tuyau , ſelon qu'il ſera ouuert diuerſement , & ainſi des autres.

Or entre ces figures , ce ne ſont pas celles qui s'impriment dans les organes des ſens exterieurs , ou dans la ſuperficie interieure du cerueau , mais ſeulement celles qui ſe tracent dans les eſprits ſur la ſuperficie de la glande H , *où eſt le ſiege de l'Imagination , & du ſens commun,* qui doiuent eſtre priſes pour les idées , c'eſt à dire pour les formes ou images que l'Ame Raiſonnable conſiderera immediatement , lors qu'eſtant vnie à cette machine elle imaginera ou ſentira quelque objet.

LXX.
Que ces impreſſiõs ſont les ſeules idées que l'ame contemplera pour ſentir ou imaginer.

Et notez que ie dis imaginera , ou ſentira ; dautant que ie veux comprendre generalement ſous le nom *d'Idée* , toutes les impreſſions , que peuuent receuoir les Eſprits en ſortant de la glande H , leſquelles s'attribüent toutes au ſens commun , lors qu'elles dépendent de la preſence des objets ; mais elles peuuent auſſi proceder de pluſieurs autres cauſes ; ainſi que ie vous diray cyaprés , & alors c'eſt à l'imagination qu'elles doiuent eſtre attribuées.

LXXI.
Qu'elle difference il y a entre ſentir , & imaginer.

Et ie pourrois adiouter icy , comment les traces de ces idées paſſent par les arteres vers le cœur , & ainſi rayonnent en tout le ſang ; Et comment meſme elles peuuent quelquefois eſtre determinées par certaines actions de la mere , à s'imprimer ſur les membres de l'enfant qui ſe forme dans ſes entrailles. Mais ie me

K

contenteray de vous dire encore, comment elles s'impriment en la partie interieure du cerueau marquée B, où est le siege de la *Memoire*.

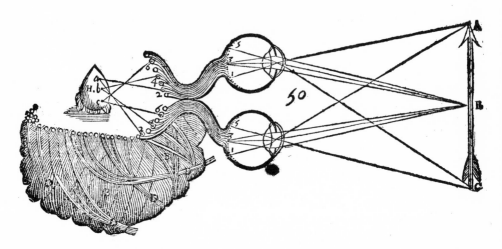

LXXII.
Comment
les traces
ou les idées
des objets
se reseruent
en la me-
moire.

Pensés donc à cet effet, qu'aprés que les esprits qui sortent de la glande H, y ont receu l'impression de quelque idée, ils passent de là par les tuyaux 2, 4, 6, & semblables, dans les pores ou interualles qui sont entre les petits filets dont cette partie du cerueau B, est composée; & qu'ils ont la force d'élargir quelque peu ces interualles, & de plier & disposer diuersement les petits filets qu'ils rencontrent en leurs chemins, selon les diuerses façons dont ils se meuuent, & les diuerses ouuertures des tuyaux par où ils passent; en sorte qu'ils y tracent aussi des figures, qui se raportent à celles des objets; non pas toutesfois si aisement ny si parfaitement du premier coup, que sur la glande H, mais peu à peu de mieux en mieux, selon que leur action est plus forte, & qu'elle dure plus long-temps, ou qu'elle est plus de fois reïterée. Ce qui eſt cauſe que ces figures ne s'effacent pas non plus ſi aiſement , mais qu'elles s'y conſeruent en telle ſorte, que par leur moyen les idées qui ont eſté autrefois ſur cette glande, s'y peuuent for-mer derechef long-temps apres, ſans que la preſence des objets auſquels elles ſe rapportent y ſoit requiſe. Et c'eſt en quoy conſiſte la *Memoire.*

Par exemple, quand l'action de l'objet A B C, aug-mentant l'ouuerture des tuyaux 2, 4, 6, eſt cauſe que les Eſprits entrent dedans en plus grande quantité qu'ils ne feroient pas ſans cela, elle eſt auſſi cauſe que paſſans plus outre vers N, ils ont la force de s'y former certains paſſages qui demeurent ouuerts, encore aprés que l'ac-tion de l'objet A B C a ceſſé; ou qui du moins s'ils ſe referment, laiſſent vne certaine diſpoſition dans les pe-tits filets dont cette partie du cerueau N eſt compoſée, par le moyen de laquelle ils peuuent beaucoup plus ai-ſement eſtre ouuerts derechef , que s'ils ne l'auoient point encore eſté; ainſi que ſi on paſſoit pluſieurs ai-guilles, ou poinçons , au trauers d'vne toile, comme vous voyez en celle qui eſt marquée A , les petits trous qu'on y feroit demeureroient encore ouuerts, comme vers a & vers b, aprés que ces aiguilles en ſeroient oſtées; ou s'ils ſe refermoient, ils laiſſeroient des traces en cette toile, comme vers c & vers d, qui ſeroient cauſe qu'on les pourroit rouurir fort aiſement.

Voyez la Figure cy-deſſous.

EXXIII.
Comment
le fouuenir
d'vne cho-
fe peut étre
excité par
vne autre. Et mefme il faut remarquer que fi on en rouuroit feulement quelques-vns , comme a & b, cela feul pourroit eftre caufe que les autres, comme c & d, fe rouuriroient auffi en mefme temps ; principalement s'ils auoient efté ouuerts plufieurs fois tous enfemble , & n'euffent pas couftume de l'eftre les vns fans les autres. Ce qui monftre comment la fouuenance d'vne chofe peut eftre excitée par celle d'vne autre , qui a efté autrefois imprimée en mefme tems qu'elle en la Memoire. Comme fi ie vois deux yeux auec vn nez, ie m'imagine auffi toft vn front & vne bouche, & toutes les autres parties d'vn vifage, pour ce que ie n'ay pas accoutumé de les voir l'vne fans l'autre ; Et voyant du feu, ie

me reſſouuiens de ſa chaleur, pour ce que ie l'ay ſentie autrefois en le voyant.

Conſiderez outre cela que la glande H eſt compoſée d'vne matiere qui eſt fort molle, & qu'elle n'eſt pas toute iointe & vnie à la ſubſtance du cerueau, mais ſeulement attachée à de petites arteres (dont les peaux ſont aſſez laſches & pliantes) & ſouſtenüe comme en balance par la force du ſang que la chaleur du cœur pouſſe vers elle; en ſorte qu'il faut fort peu de choſe pour la determiner à s'incliner & ſe pancher plus ou moins, tantoſt d'vn coſté tantoſt d'vn autre, & faire qu'en ſe panchant, elle diſpoſe les eſprits qui ſortent d'elle, à prendre leur cours vers certains endroits du cerueau, plutoſt que vers les autres.

LXXIV. Qu'il faut fort peu de choſe pour determiner la glande à s'incliner d'vn coſté ou d'autre.

Or il y a deux cauſes principales, ſans conter la force de l'Ame, que ie mettray cy-apres, qui la peuuent ainſi faire mouuoir, & qu'il faut icy que ie vous explique.

La premiere eſt la difference qui ſe rencontre entre les petites parties des Eſprits qui ſortent d'elle : Car ſi tous ces Eſprits eſtoient exactement d'égale force, & qu'il n'y euſt aucune autre cauſe qui la determinaſt à ſe pancher n'y çà ny là, ils couleroient également dans tous ſes pores, & la ſoutiendroient toute droite & immobile au centre de la teſte, ainſi qu'elle eſt repreſentée en la figure 40. Mais comme vn corps attaché ſeulement à quelques filets, qui ſeroit ſoutenu en l'air par la force de la fu-

LXXV. Que la difference qui eſt entre les eſprits, eſt l'vne des cauſes qui la determinent.

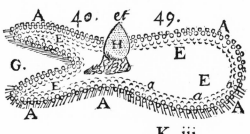

40. et 49.

K iij

mée qui fortiroit d'vn fourneau , flotteroit inceſſam-
ment çà & là, ſelon que les diuerſes parties de cette fumée agiroient contre luy diuerſement ; Ainſi les petites parties de ces Eſprits , qui ſouleuent & ſoutiennent
cette glande eſtans preſque touſiours differentes en
quelque choſe , ne manquent pas de l'agiter & faire
pancher tantoſt d'vn coſté tantoſt d'vn autre , comme

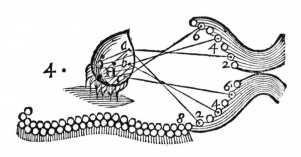

4.

vous la voyez en cette figure 41 , où non ſeulement ſon
centre H eſt vn peu éloigné du centre du cerueau marqué o , mais auſſi les extremitez des arteres qui la ſoutiennent, ſont courbées en telle ſorte , que preſque tous
les Eſprits qu'elles luy apportent, prennent leur cours
par l'endroit de ſa ſuperficie a, b, c, vers les petits tuyaux 2, 4, 6, ouurans par ce moyen ceux de ſes pores qui
regardent vers là, beaucoup dauantage que les autres.

LXXVI.
Quel eſt le
principal
effet des eſ-
prits qui
ſortent de
la glande.
          Or le principal effet qui ſuit de cecy , conſiſte , en ce
que les Eſprits ſortans ainſi plus particulierement de
quelques endroits de la ſuperficie de cette glande , que
des autres , peuuent auoir la force de tourner les petits
tuyaux de la ſuperficie interieure du cerueau dans leſquels ils ſe vont rendre, vers les endroits d'où ils ſortent,
s'ils ne les y trouuent deſia tout tournez ; & par ce moyen

de faire mouuoir les membres auſquels ſe raportent ces tuyaux, vers les lieux auſquels ſe raportent ces endroits de la ſuperficie de la glande **H**. Et notez que l'idée de ce mouuement des membres ne conſiſte qu'en la façon dont ces Eſprits ſortent pour lors de cette glande, & ainſi que c'eſt ſon idée qui le cauſe.

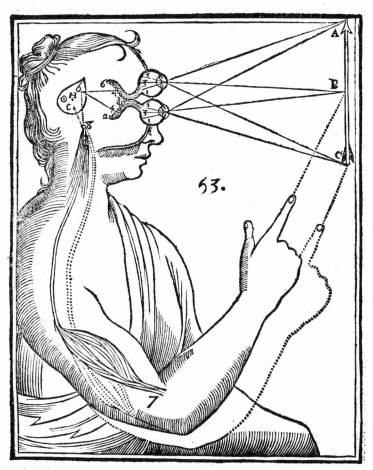

Comme icy par exemple, on peut ſuppoſer, que ce qui fait que le tuyau 8, ſe tourne plutoſt vers le point b, que

LXXVII. En quoy conſiſte

l'idée du
mouue-
ment des
membres;
& que ſa
ſeule idée le
peut cau-
ſer.

vers quelqu'autre , c'eſt ſeulement que les Eſprits qui
ſortent de ce point, tendent auec plus de force vers luy
qu'aucuns autres ; & que cela meſme donneroit occaſion à l'Ame, de ſentir que le bras ſe tourne vers l'objet
B, ſi elle eſtoit deſia dans cette machine, ainſi que ie l'y
ſuppoſeray cy-aprés. Car il faut penſer que tous les
points de la glande vers leſquels ce tuyau 8 peut eſtre
tourné, répondent tellement à tous les lieux vers leſquels le bras marqué 7 le peut eſtre , que ce qui fait
maintenant que ce bras eſt tourné vers l'objet B, c'eſt
que ce tuyau regarde le point b de la glande ; Que ſi
les Eſprits changeans leur cours tournoient ce tuyau
vers quelqu'autre point de la glande , comme vers c,
les petits filets 8, 7, qui ſortans d'autour de luy ſe vont
rendre dans les muſcles de ce bras, changeans par meſme moyen de ſituation, retreciroient quelques-vns des
pores du cerueau qui ſont vers D, & en élargiroient
quelques autres : Ce qui feroit que les Eſprits, paſſans
de là dans ces muſcles d'autre façon qu'ils ne font à pre-
ſent, tourneroient incontinent ce bras vers l'objet C;
comme reciproquement , ſi quelqu'autre action que
celle des Eſprits qui entrent par le tuyau 8, tournoit ce
meſme bras vers B ou vers C, elle feroit que ce tuyau
8 ſe tourneroit vers les points de la glande b ou c; en
ſorte que l'idée de ce mouuement ſe formeroit auſſi en
meſme temps, au moins ſi l'attention n'en eſtoit point
diuertie, c'eſt à dire, ſi la glande H n'eſtoit point empeſchée de ſe pancher vers 8, par quelqu'autre action
qui fuſt plus forte. Et ainſi generalement il faut penſer,
que chacun des autres petits tuyaux qui ſont en la ſuperficie

perficie interieure du cerueau, ſe raporte à chacun des
autres membres; & chacun des autres points de la ſuperficie de la glande H à chacun des coſtez vers leſquels
ces membres peuuent eſtre tournez : En ſorte que les
mouuemens de ces membres, & leurs idées, peuuent
eſtre cauſez reciproquement l'vn par l'autre.

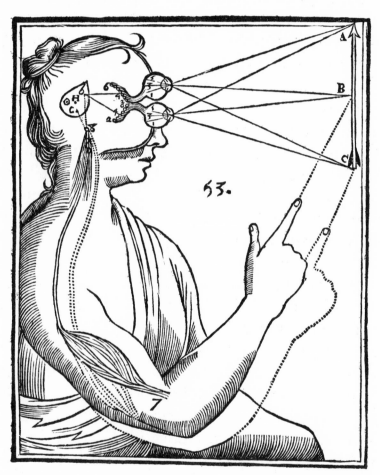

Et de plus, pour entendre icy par occaſion, comment, LXXVIII.
lors que les deux yeux de cette machine, & les organes Comment
                                                        vne idée
L

peut-eſtre
cõpoſée de
pluſieurs ;
& d'où viẽt
qu'alors il
ne paroiſt
qu'vn ſeul
objet.

de pluſieurs autres de ſes ſens, ſont tournez vers vn meſme objet, il ne s'en forme pas pour cela pluſieurs idées
dans ſõn cerueau, mais vne ſeule, il faut penſer que c'eſt
touſiours des meſmes points de cette ſuperficie de la
glande H que ſortent les eſprits, qui tendans vers diuers tuyaux peuuent tourner diuers membres vers les
meſmes objets : Comme icy que c'eſt du ſeul point b
que ſortent les eſprits qui tendans vers les tuyaux 4,4,
& 8, tournent en meſme temps les deux yeux & le bras
droit vers l'objet B.

LXXIX.
En quoy
conſiſte l'i-
dée de la
diſtance
des objets.

Ce qui vous ſera facile à croire, ſi pour entendre auſſi
en quoy conſiſte l'idée de la diſtance des objets, vous
penſez que ſelon que cette ſuperficie change de ſituation, les meſmes de ſes points ſe raportent à des lieux,
dautant plus éloignez du centre du cerueau marqué o,
que ces points en ſont plus proches, & dautant plus
proches qu'ils en ſont plus éloignez. Comme icy il faut
penſer que ſi le point b, eſtoit vn peu plus retiré en arriere qu'il n'eſt pas, il ſe raporteroit à vn lieu plus éloigné que n'eſt B, & s'il eſtoit vn peu plus panché en a-
uant, il ſe raporteroit à vn plus proche.

LXXX.
Que la di-
uerſe ſitua-
tion de la
glandepeut
faire ſentir
diuers ob-
jets, ſans
aucũ chan-
gement
dans l'or-
gane.

Et cecy ſera cauſe, que lors qu'il y aura vne Ame dans
cette machine, elle pourra quelquefois ſentir diuers
objets par l'entremiſe des meſmes organes, diſpoſez en
meſme ſorte, & ſans qu'il y ait rien du tout qui ſe change, que la ſituation de la glande H. Comme icy par
exemple, l'Ame pourra ſentir ce qui eſt au point L, par
l'entremiſe des deux mains, qui tiennẽt les deux baſtons
N L & O L pour ce que c'eſt du point L, de la glande H,
que ſortent les Eſprits qui entrent dans les tuyaux 7, & 8,

aufquels répondent fes deux mains ; au lieu que fi cette
glande H, eſtoit vn peu plus en auant qu'elle n'eſt, en
forte que les points de ſa ſuperficie n & o, fuſſent aux
lieux marquez i, & k, & par conſequent que ce fuſt
d'eux, que ſortiſſent les Eſprits qui vont vers 7 & vers 8,
l'Ame deuroit ſentir ce qui eſt vers N, & vers O par l'entremiſe des meſmes mains , & ſans qu'elles fuſſent en
rien changées.

L ij

LXXXI.
Que les
veltiges de
la memoi-
re font auf-
fi vne des
caufes qui
font pan-
cher la
glande.

Au refte, il faut remarquer que lors que la glande H eft panchée vers quelque cofté, par la feule force des Efprits, & fans que l'Ame Raifonnable, ny les fens exterieurs y contribuent, les idées qui fe forment fur fa fuperficie ne procedent pas feulement des inegalitez, qui fe rencontrent entre les petites parties de ces Efprits, & & qui caufent la difference des humeurs, ainfi qu'il a efté dit cy-deffus, mais elles procedent auffi des impreffions de la Memoire. Car fi la figure de quelque objet particulier eft imprimée beaucoup plus diftinctement qu'aucune autre, à l'endroit du cerueau vers lequel eft iuftement panché cette glande, les Efprits qui tendent vers là ne peuuent manquer d'en receuoir auffi l'impreffion. Et c'eft ainfi que les chofes paffées reuiennent quelquefois en la penfée, comme par hazard, & fans que la Memoire en foit fort excitée par aucun objet qui touche les fens.

LXXXII.
Comment
fe forment
les fantof-
mes en l'i-
maginatiõ
de ceux qui
réuẽt étant
éueillez.

Mais fi plufieurs diuerfes figures fe trouuent tracées en ce mefme endroit du cerueau, prefqu'auffi parfaitement l'vne que l'autre, ainfi qu'il arriue le plus fouuent, les Efprits receuront quelque chofe de l'impreffion de chacune, & ce plus ou moins felon la diuerfe rencontre de leurs parties; Et c'eft ainfi que fe compofent les chymeres, & les hypogrifes, en l'imagination de ceux qui réuent eftant éueillez, c'eft à dire qui laiffent errer nonchalamment çà, & là leur fantaifie, fans que les objets exterieurs la diuertiffent, n'y qu'elle foit conduite par leur raifon.

LXXXIII
Que cette
machine

Mais l'effet de la Memoire qui me femble icy le plus digne d'eftre confideré, confifte, en ce que fans qu'il y

ait aucune Ame dans cette machine, elle peut naturellement eſtre diſpoſée, à imiter tous les mouuemens que
de vrais hommes, ou bien d'autres ſemblables machines, feront en ſa preſence.

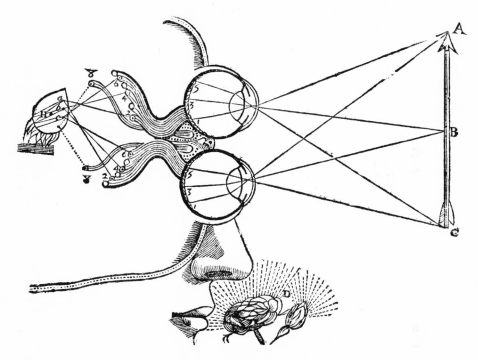

La ſeconde cauſe qui peut determiner les mouuemens
de la glande H, eſt l'action des objets qui touchent les
ſens. Car il eſt aiſé à entendre, que l'ouuerture des petits tuyaux 2, 4, 6, par exemple, eſtant élargie par l'action de l'objet A B C, les Eſprits qui commencent auſſi-
toſt à couler vers eux, plus librement & plus viſte qu'ils
ne faiſoient, attirent apres ſoy quelque peu cette glande, & font qu'elle ſe panche, ſi elle n'en eſt d'ailleurs
empeſchée; & changeans la diſpoſition de ſes pores, elle

commence à conduire beaucoup plus grande quantité
d'Efprits par a, b, c, vers 2, 4, 6, qu'elle ne faiſoit auparauant ; ce qui rend l'idée que forment ces Efprits
d'autant plus parfaite. Et c'eſt en quoy conſiſte le premier effet, que ie defire que vous remarquiez.

LXXXV.
Que les di-
uerſes idées
qui s'im-
primēt ſur
la glande
s'empeſ-
chent l'vne
l'autre.

Le ſecond conſiſte, en ce que pendant que cette glande eſt retenüe ainſi panchée vers quelque coſté, cela
l'empeſche de pouuoir ſi aiſement receuoir les idées des
objets qui agiſſent contre les organes des autres ſens.
Comme icy par exemple, pendant que preſque tous les
eſprits que produit la glande H, ſortent des points à,
b, c, il n'en ſort pas aſſez du point d, pour y former l'i-
dée de l'objet D, dont ie ſuppoſe que l'action n'eſt ny
ſi viue, ny ſi forte, que celle d'A, B, C ; D'où vous voyez
comment les idées s'empeſchent l'vne l'autre, & d'où
vient qu'on ne peut eſtre fort attentif à pluſieurs choſes
en meſme temps.

LXXXVI.
Que la
preſence
d'vn obiet
ſuffit pour
diſpoſer
l'œil à en
bien rece-
uoir l'ac-
tion.

Il faut auſſi remarquer, que les organes des ſens, lors
qu'ils commencent à eſtre touchez par quelque objet
plus fort que par les autres, n'eſtans pas encore autant
diſpoſez à en receuoir l'action qu'ils pourroient eſtre,
la preſence de cet objet eſt ſuffiſante pour acheuer de les
y diſpoſer entierement. Comme ſi l'œil par exemple,
eſt diſpoſé à regarder vers vn lieu fort éloigné, lors que
l'objet A B C, qui eſt fort proche, commence à ſe pre-
ſenter deuant luy, ie dis que l'action de cet objet pourra faire qu'il ſe diſpoſera tout auſſi-toſt à le regarder
fixement.

LXXXVII.
Quelle dif-
ference il y
a entre

Et afin que cecy vous ſoit plus aiſé à entendre, conſiderez premierement la difference qui eſt entre l'œil,

difpofé à regarder vn objet éloigné, comme il eft en la
50ᵉ figure p. 74. & le mefme œil, difpofé à en regarder vn
plus proche, comme il eft en cette 51. qui confifte, non
feulement en ce que l'humeur cryftalline eft vn peu plus
voûtée, & les autres parties de l'œil à proportion autrement difpofées en cette derniere figure qu'en la precedente, mais auffi en ce que les petits tuyaux 2, 4, 6, y

l'œil difpo-
fé à regar-
der vn ob-
jet proche,
ou vn éloi-
gné.

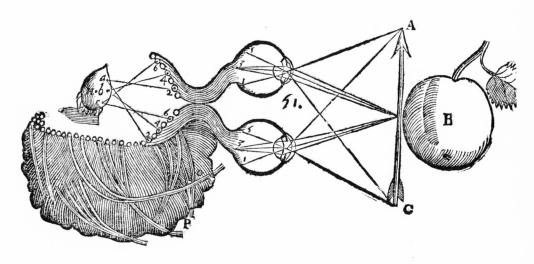

font inclinez vers vn point plus proche, & que la glande
H y eft vn peu plus auancée vers eux, & que l'endroit de
fa fuperficie a, b, c, y eft à proportion vn peu plus voûté
ou courbé; en forte qu'en l'vne & en l'autre figure, c'eft
toufiours du point a, que fortent les Efprits qui tendent
vers le tuyau 2; du point b, que fortent ceux qui tendent vers le tuyau 4; & du point c, que fortent ceux qui
tendent vers le tuyau 6.

Confiderez auffi que les feuls mouuemens de la glande H, font fuffifans pour changer la fituation de ces

tuyaux, & en fuitte toute la difpofition du corps de l'œil, ainfi qu'il a tantoft efté dit en general qu'ils peuuent faire mouuoir tous les membres.

LXXXVIII.
Que les
pores du
cerueau
peuuent
eftre dau-
tant plus
ouuerts,que
l'œil eft
mieux dif-
pofé à re-
ceuoir l'ac-
tion de fon
objet.
Confiderez apres cela que ces tuyaux 2, 4, 6, peuuent eftre dautant plus ouuerts par l'action de l'objet A B C, que l'œil eft plus difpofé à le regarder : Car fi les rayons qui tombent fur le point 3 par exemple, viennent tous du point B, comme ils font lors que l'œil regarde fixe-ment vers là, il eft euident que leurs actions doiuent tirer plus fort le petit filet 34, que s'ils venoient partie du point A, partie de B, & partie de C, comme ils font fi toft que l'œil eft vn peu autrement difpofé ; à caufe que pour lors leurs actions n'eftant pas fi femblables, ny fi vnies, ne peuuent eftre du tout fi fortes, & s'empef-chent mefme fouuent l'vne l'autre; Ce qui n'a lieu neant-moins que touchant les objets dont les lineamens ne font ny trop femblables ny trop confus; comme auffi n'y a-t'il que ceux là dont l'œil puiffe bien diftinguer la diftance, & difcerner les parties, ainfi que i'ay remar-qué en la Dioptrique.

LXXXIX.
Que la glä-
de fe pan-
che plus ai-
fémét vers
le cofté qui
fert à mieux
difpofer
l'œil.
De plus confiderez que la glande H, peut beaucoup plus facilement eftre meüe, vers le cofté vers lequel en fe panchant elle difpofera l'œil à receuoir plus diftin-ctement qu'il ne fait l'action de l'objet qui agit le plus fort de tous contre luy, que vers ceux où elle pourroit faire le contraire. Comme par exemple, en cette 50 fi-gure, où l'œil eft difpofé à regarder vn objet éloigné, il faut bien moins de force pour l'inciter à fe pancher vn peu plus en auant qu'elle n'eft, que pour faire qu'elle fe retire plus en arriere ; pource qu'en fe retirant elle
rendroit

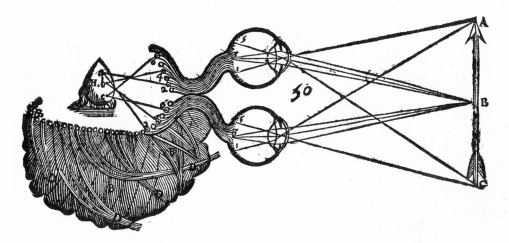

rendroit l'œil encore moins difpofé qu'il n'eft pas, à receuoir l'action de l'objet A B C, que l'on fuppofe eftre
proche, & agir le plus fort de tous contre luy, & ainfi
elle feroit caufe que les petits tuyaux 2, 4, 6, feroient
auffi moins ouuerts par cette action, & que les Efprits
qui fortent des points a, b, c, couleroient auffi moins
librement vers ces tuyaux; Au lieu qu'en s'auançant, elle feroit tout au contraire que l'œil fe difpofant mieux
à receuoir cette action, les petits tuyaux 2, 4, 6, s'ouuri-
roient dauantage, & en fuite que les Efprits qui fortent
des points a, b, c, couleroient vers eux plus librement;
En forte mefme que fi-toft que la glade auroit le moins
du monde commencé ainfi à fe mouuoir, le cours
de ces Efprits l'emporteroit tout auffi-toft, & ne luy
permettroit pas de s'arrefter, iufqu'à ce qu'elle fuft tout
à fait difpofée en la façon que vous la voyez en la 51 figure, & que l'œil regardaft fixement vers cet objet proche A B C.

M

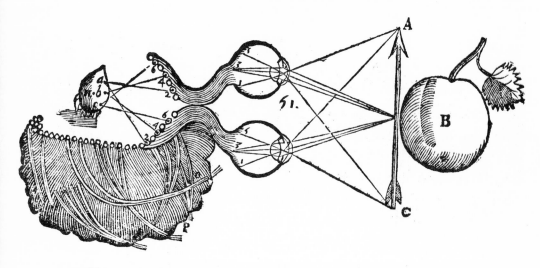

XC.
Qu'eſt-ce
qui com-
mence or-
dinaire-
mēt à faire
mouuoir
& incliner
la glande
quelque
part.

Si bien qu'il ne reſte plus qu'à vous dire la cauſe qui peut commencer ainſi à la mouuoir; laquelle n'eſt autre ordinairemēt que la force de l'objet meſme, qui agiſſant contre l'organe de quelque ſens, augmente l'ouuerture de quelques-vns des petits tuyaux qui ſont en la ſuperficie interieure du cerueau, vers leſquels les Eſprits commençans auſſi-toſt à prendre leurs cours, attirent auec ſoy cette glande, & la font incliner vers ce coſté là. Mais en cas que ces tuyaux fuſſent deſia d'ailleurs autant ou plus ouuerts que cet objet ne les ouure, il faut penſer que les petites parties des Eſprits qui coulent au trauers de ſes pores, eſtant inegales, la pouſſent tantoſt deçà tantoſt de là, fort promptement, & en moins d'vn clin d'œil de tous coſtez, ſans la laiſſer iamais en repos vn ſeul moment; & que s'il ſe rencontre d'abord, qu'elles la pouſſent vers vn coſté, vers lequel il ne luy ſoit pas aiſé de s'incliner, leur action, qui n'eſt

pas de foy grandement forte, ne peut prefque auoir aucun effet ; mais au contraire fi toft qu'elles la pouffent le
moins du monde, vers le cofté vers lequel elle eft defia
toute portée, elle ne manquera pas de s'incliner verslà
auffi-toft, & en fuite de difpofer l'organe du fens à receuoir l'action de fon objet, le plus parfaitement qu'il eft
poffible, ainfi que ie viens d'expliquer.

Acheuons maintenant de conduire les Efprits iufques
aux nerfs, & voyons les mouuemens qui en dependent.
Si les petits tuyaux de la fuperficie interieure du cerueau
ne font point du tout plus ouuerts, ny d'autre façon, les
vns que les autres, & par confequent que ces Efprits
n'ayent en eux l'impreffion d'aucune idée particuliere,

xci.
Comment
les Efprits
font con-
duits dans
les nerfs,
pour mou-
uoir cette
machine.

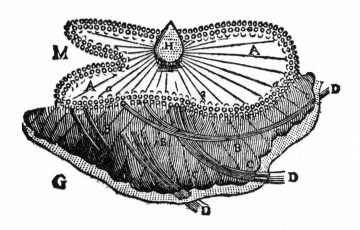

ils fe répandent indifferemment de tous coftez, & paffent des pores qui font vers B, en ceux qui font vers C,
d'où les plus fubtiles de leurs parties s'écouleront tout à
fait hors du cerueau, par les pores de la petite peau qui
l'enuelope ; puis le furplus prenant fon cours vers D, s'ira

M ij

rendre dans les nerfs, & dans les mufcles, fans y caufer aucun effet particulier, pource qu'il fe diftribuera en tous également.

Mais s'il y a quelques-vns des tuyaux qui foient plus ou moins ouuerts, ou feulement ouuerts de quelqu'autre façon que leurs voifins, par l'action des objets qui meuuent les fens, les petits filets qui compofent la fubftance du cerueau, eftans en fuite vn peu plus tendus ou plus lafches les vns que les autres, conduiront les Efprits vers certains endroits de fa bafe, & de là vers certains nerfs, auec plus ou moins de force que vers les autres; Ce qui fuffira pour caufer diuers mouuemens dans les mufcles, fuiuant ce qui a efté cy-deffus amplement expliqué.

XCII.
De fix diuerfes circonftances d'où peuuent dependre fes mouuemens.

Or d'autant que ie veux vous faire conceuoir ces mouuemens, femblables à ceux aufquels nous fommes naturellement incitez par les diuerfes actions des objets qui meuuent nos fens, ie defire icy que vous confideriez fix diuerfes fortes de circonftances dont ils peuuent dependre. La premiere eft le lieu d'où procede l'action qui ouure quelques-vns des petits tuyaux par où entrent premierement les Efprits. La feconde confifte en la force & en toutes les autres qualitez de cette action. La troifiéme, en la difpofition des petits filets qui compofent la fubftance du cerueau. La quatriéme, en l'inegale force que peuuent auoir les petites parties des Efprits. La cinquiéme, en la diuerfe fituation des membres exterieurs. Et la fixiéme en la rencontre de plufieurs actions qui meuuent les fens en mefme temps.

XCIII. Pour le lieu d'où procede l'action, vous fçauez defia,

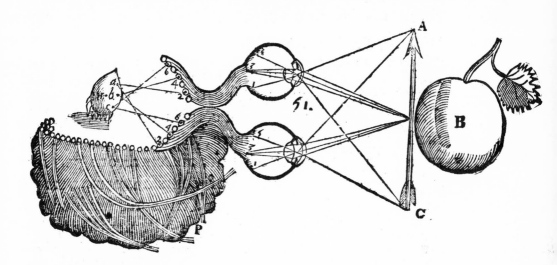

que fi l'objet A B C, par exemple, agiffoit contre vn autre fens, que contre celuy de la veüe, il ouuriroit d'autres tuyaux, en la fuperficie interieure du cerueau, que ceux qui font marquez 2, 4, 6; Et que s'il eftoit plus prés, ou plus loin, ou autrement fitué au refpect de l'œil qu'il n'eft pas, il pourroit bien à la verité ouurir ces mefmes tuyaux, mais qu'il faudroit qu'ils fuffent autrement fituez qu'ils ne font, & par confequent qu'ils puffent receuoir des Efprits d'autres points de la glande que de ceux qui font marquez a, b, c, & les conduire vers d'autres endroits que vers A B C, où ils les conduifent maintenant, & ainfi des autres.

Pour les diuerfes qualitez de l'action qui ouure ces tuyaux, vous fçauez auffi que felon qu'elles font differentes, elle les ouure diuerfement; & il faut penfer que cela feul eft fuffifant, pour changer le cours des Efprits dans le cerueau. Comme par exemple, fi l'objet A B C

eſt rouge, c'eſt à dire, s'il agit contre l'œil 1, 3, 5, en la
façon que i'ay dit cy-deſſus eſtre requiſe pour faire ſentir la couleur rouge, & qu'auec cela il ait la figure d'vne
pomme, ou autre fruit, il faut penſer qu'il ouurira les
tuyaux 2, 4, 6, d'vne certaine façon particuliere, qui
ſera cauſe que les parties du cerueau qui ſont vers N, ſe
preſſeront l'vne contre l'autre, vn peu plus que de coutume ; en ſorte que les Eſprits qui entreront par ces
tuyaux 2, 4, 6, prendront leur cours d'N par o vers p.
Et que ſi cet objet A B C eſtoit d'vne autre couleur, ou
d'vne autre figure, ce ne ſeroit pas iuſtement les petits
filets qui ſont vers N & vers o qui detourneroient les
Eſprits qui entrent par 2, 4, 6, mais quelques autres de
leurs voiſins.

Et ſi la chaleur du feu A, qui eſt proche de la main B,
n'eſtoit que mediocre, il faudroit penſer, que la façon
dont elle ouuriroit les tuyaux 7, ſeroit cauſe que les parties du cerueau qui ſont vers N, ſe preſſeroient, & que
celles qui ſont vers o, s'élargiroient vn peu plus que de
coutume ; & ainſi que les Eſprits qui viennent du tuyau
7, iroient d'N par o vers p. Mais ſuppoſant que ce feu
brûle la main, il faut penſer que ſon action ouure tant
ces tuyaux 7, que les Eſprits qui entrent dedans, ont la
force de paſſer plus loin en ſigne droite que iuſques à N,
à ſçauoir iuſques à o & à R, où pouſſant deuant eux les
parties du cerueau qui ſe trouuent en leur chemin, ils les
preſſent en telle ſorte, qu'ils ſont repouſſez & detournez
par elles vers S, & ainſi des autres.

XCV.
La 3. eſt la
diſpoſition
Pour la diſpoſition des petits filets qui compoſent la
ſubſtance du cerueau, elle eſt ou Acquiſe, ou Naturelle,

Et pource que l'Acquife eft dependante de toutes les au- naturelle
tres circonftances qui changent le cours des Efprits, ie ou aequife
la pourray tantoft mieux expliquer. Mais afin que ie des petits
vous die en quoy confifte la Naturelle ; fçachez que la fubftan-
Dieu a tellement difpofé ces petits filets en les formant, ce du cer-
que les paffages qu'il a laiffez parmy eux, peuuent con- ueau.
duire les Efprits, qui font meus par quelque action par-

ticuliere, vers tous les nerfs où ils doiuent aller, pour
cauſer les meſmes mouuemens en cette machine, auſquels vne pareille action nous pourroit inciter, ſuiuant
les inſtincts de noſtre nature; En ſorte qu'icy par exemple, où le feu A brûle la main B, & eſt cauſe que les
Eſprits qui entrent dans le tuyau 7 tendent vers o, ces
Eſprits trouuent là deux pores, ou paſſages principaux
o R, o s; l'vn deſquels, à ſçauoir o R, les conduit en
tous les nerfs qui ſeruent à mouuoir les membres exterieurs, en la façon qui eſt requiſe pour euiter la force
de cette action, comme en ceux qui retirent la main,
ou le bras, ou tout le corps; & en ceux qui tournent la
teſte & les yeux vers ce feu, afin de voir plus particulierement ce qu'il faut faire pour s'en garder. Et par l'autre o s, ils vont en tous ceux qui ſeruent à cauſer des
émotions interieures, ſemblables à celles qui ſuiuent
en nous de la douleur; comme en ceux qui reſſerrent le
cœur, qui agitent le foye, & tels autres; Et meſme auſſi
en ceux qui peuuent cauſer les mouuemens exterieurs
qui la témoignent; Comme en ceux qui excitent les
larmes, qui rident le front & les iouës, & qui diſpoſent
la voix à crier. Au lieu que ſi la main B, eſtant fort froi-
de, le feu A la réchauffoit moderement, & ſans la bru-
ler, il ſeroit cauſe que les meſmes Eſprits, qui entrent
par le tuyau 7, iroient ſe rendre non plus vers O & vers
R, mais vers o & vers p, où ils trouueroient derechef
des pores, diſpoſez à les conduire en tous les nerfs qui
peuuent ſeruir aux mouuemens conuenables à cette action.

Et

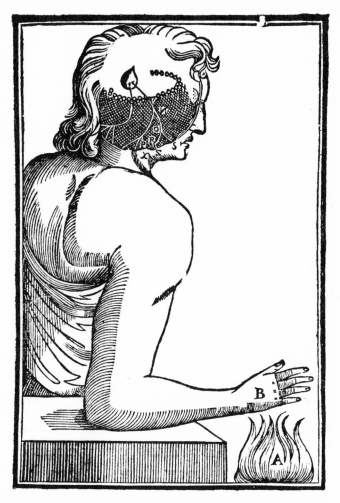

Et remarquez que i'ay particulierement diſtingué les deux pores o R & o s, pour vous aduertir, qu'il y a preſque touſiours deux ſortes de mouuemens qui procedent de chaque action ; ſçauoir les exterieurs, qui ſeruent à pourſuiure les choſes deſirables, ou à euiter les nuiſibles, & les interieurs, qu'on nomme commune-ment *les paſſions*, qui ſeruent à diſpoſer le cœur & le

XCVI.
Qu'il y a preſque touſiours deux ſortes de mouue-mens qui procedent de chaque action.

N

foye, & tous les autres organes defquels le temperament du fang, & en fuite celuy des Efprits peut dependre; en telle forte que les Efprits qui naiffent pour lors,
fe trouuent propres à caufer les mouuemens exterieurs
qui doiuent fuiure. Car fuppofant que les diuerfes qualitez de ces Efprits font l'vne des circonftances qui feruent à changer leur cours, ainfi que i'expliqueray tout
maintenant, on peut bien penfer, que fi par exemple,
il eft queftion d'éuiter quelque mal par la force, & en le
furmontant, ou le chaffant, à quoy incline la paffion
de la *colere*, les Efprits doiuent eftre plus inégalement
agitez, & plus forts que de coutume; Et au contraire,
que s'il faut l'euiter, en fe cachant, ou le fupporter auec
patience, à quoy incline la paffion de la *peur*, ils doiuent
eftre moins abondans, & moins forts; Et pour cet effet
le cœur fe doit refferrer pour lors, comme pour les épar-
gner & referuer pour le befoin : Et vous pouuez iuger
des autres paffions à proportion.

Quant aux autres mouuemens exterieurs, qui ne feruent point à éuiter le mal, ou à fuiure le bien, mais feulement à témoigner les paffions, comme ceux en quoy
confifte le rire, ou le pleurer, ils ne fe font que par occafion, & pource que les nerfs par où doiuent entrer les
Efprits pour les caufer, ont leur origine tout proche de
ceux par où ils entrent pour caufer les paffions, ainfi
que l'Anatomie vous peut apprendre.

XCVII.
La 4. eft
l'inegale
force des
efprits; &
comment
elle peut

Mais ie ne vous ay pas encore fait voir, comment les
diuerfes qualitez des Efprits peuuent auoir la force de
changer la determination de leur cours; Ce qui arriue
principalement lors que d'ailleurs ils ne font que fort

peu ou point du tout determinez. Comme fi les nerfs de changer la determina-tiõ de leur cours.
l'eftomac font agitez en la façon que i'ay dit cy-deffus
qu'ils doiuent eftre, pour caufer le fentiment de la faim,
& que cependant il ne fe prefente rien à aucun fens, ny
à la memoire, qui paroiffe propre à eftre mangé, les Efprits que cette action fera entrer par les tuyaux 8. dans
le cerueau, s'iront rendre en vn endroit, où ils trouue-
ront plufieurs pores difpofez à les conduire indifferemment en tous les nerfs qui peuuent feruir à la recherche,
ou à la pourfuitte de quelqu'objet; en forte qu'il n'y aura que la feule inegalité de leurs parties, qui puiffe eftre
caufe qu'ils prennent leur cours plutoft par les vns que
par les autres.

Et s'il arriue que les plus fortes de ces parties foient XCVIII. Comment cette ma-chine peut fembler he-fiter dans fes actions.
maintenant celles qui tendent à couler vers certains
nerfs, puis incontinent apres que ce foient celles qui
tendent vers leurs contraires, cela fera imiter à cette Machine les mouuemens qui fe voyent en nous, lors
que nous hefitons, & fommes en doute de quelque
chofe.

Tout de mefme fi l'action du feu A eft moyenne entre celles qui peuuent conduire les Efprits vers R, & vers
p, c'eft à dire entre celles qui caufent la douleur, & le
plaifir, il eft aifé à entendre que les feules inegalitez qui
font en eux, doiuent fuffire pour les determiner à l'vn
ou à l'autre; ainfi que fouuent vne mefme action, qui
nous eft agreable lors que nous fommes en bonne humeur, nous peut déplaire lors que nous fommes triftes
& chagrins. Et vous pouuez tirer de cecy la raifon de
tout ce que i'ay dit cy-deffus, touchant les humeurs ou

inclinations tant naturelles qu'acquifes, qui dependent
de la difference des Efprits.

XCIX.
La 5. eft la
diuerfe fi-
tuation des
membres
exterieurs.

Pour la diuerfe fituation des membres exterieurs, il
faut feulement penfer qu'elle change les pores qui portent immediatement ces Efprits dans les nerfs; En forte que par exemple, fi lors que le feu A brûle la main
B, la tefte eftoit tournée vers le cofté gauche, au lieu
qu'elle l'eft maintenant vers le droit, les Efprits iroient
tout de mefme qu'ils font de 7 vers N, puis vers o, & de
là vers R & vers s; Mais que de R, au lieu d'aller vers x,
par où ie fuppofe qu'ils doiuent paffer, pour redreffer la

Voyez la
fig. p. 97.

tefte qui eft tournée vers la main droite, ils iroient vers
z, par où ie fuppofe qu'ils deuroient entrer pour la redreffer, fi elle eftoit tournée vers la gauche; dautant
que la fituation de cette tefte, qui eft maintenant caufe que les petits filets de la fubftance du cerueau qui font
vers x, font beaucoup plus lafches & aifez à écarter
l'vn de l'autre, que ceux qui font vers z, eftant changée, feroit tout au contraire, que ceux qui font vers z,
feroient fort lafches, & ceux qui font vers x, fort tendus
& refferrez.

C.
Comment
cette ma-
chine mar-
the.

Ainfi pour entendre comment vne feule action, fans
fe changer, peut mouuoir maintenant vn pié de cette
Machine, maintenant l'autre, felon qu'il eft requis pour
faire qu'elle marche, il fuffit de penfer que les Efprits
paffent par vn feul pore, dont l'extremité eft autrement
difpofée, & les conduit en d'autres nerfs, quand c'eft le
pié gauche qui eft le plus auancé, que quand c'eft le
droit. Et on peut rapporter icy tout ce que i'ay dit cy-
deffus de la refpiration, & de tels autres mouuemens,

qui ne dependent ordinairement d'aucune idée; ie dis
ordinairement, car ils en peuuent quelquefois auſſi dependre.

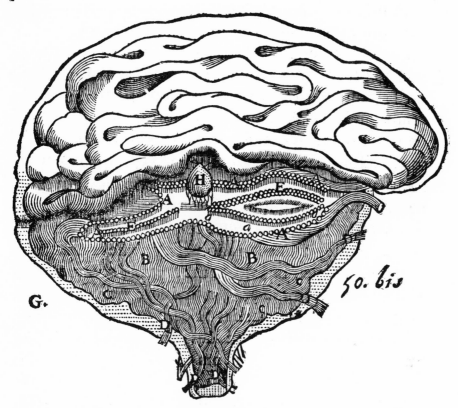

Maintenant que ie penſe auoir ſuffiſamment expliqué
toutes les fonctions de la veille, il ne me reſte que fort
peu de choſes à vous dire touchant *le ſommeil;* car premierement il ne faut que ietter les yeux ſur cette 50. figure, & voir comment les petits filets D, D, qui ſe vont
rendre dans les nerfs y ſont laſches & preſſez, pour entendre comment, lors que cette Machine repreſente le
corps d'vn homme qui dort, les actions des objets ex-

C I.
Du ſom-
meil ; & en
quoy il dif-
fere de la
veille.

N iij

terieurs font pour la plus-part empefchées de paffer iufqu'à fon cerueau, pour y eftre fenties ; & les Efprits qui
font dans le cerueau empefchez de paffer iufques aux
membres exterieurs, pour les mouuoir ; qui font les
deux principaux effets du fommeil.

CII.
Des fon-
ges ; & en
quoy ils
differét des
réueries de
la veille.

Pour ce qui eft *des fonges*, ils dependent en partie de
l'inegale force que peuuent auoir les Efprits qui fortent
de la glande H, & en partie des impreffions qui fe rencontrent dans la Memoire ; En forte qu'ils ne different
en rien, de ces idées que i'ay dit cy-deffus fe former
quelquefois dans l'imagination de ceux qui réuent é-
tant éueillez, fi ce n'eft en ce que les images qui fe forment pendant le fommeil, peuuent eftre beaucoup plus
diftinctes, & plus viues, que celles qui fe forment pendant la veille ; Dont la raifon eft, qu'vne mefme force
peut ouurir dauantage les petits tuyaux, comme 2, 4, 6,
& les pores, comme a, b, c, qui feruent à former ces
images, lors que les parties du cerueau qui les enuironnent font lafches & detendües, ainfi que vous le voyez
en cette 50. figure, que lors qu'elles font toutes ten-
dües, ainfi que vous le pouuez voir en celles qui la precedent. Et cette mefme raifon monftre auffi, que s'il arriue que l'action de quelque objet qui touche les fens,
puiffe paffer iufqu'au cerueau pendant le fommeil, elle
n'y formera pas la mefme idée qu'elle feroit pendant la
veille, mais quelqu'autre plus remarquable, & plus fen-
fible ; Comme quelquefois quand nous dormons, fi nous
fommes piquez par vne mouche, nous fongeons qu'on
nous donne vn coup d'efpée ; fi nous ne fommes pas du
tout affez couuers, nous nous imaginons eftre tout

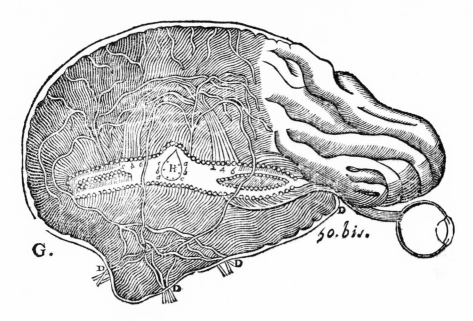

G.   50.bis.

nuds; & si nous le sommes quelque peu trop, nous pen-
sons estre accablez d'vne montagne.

Au reste, pendant le sommeil, la substance du cerueau qui est en repos, a le loisir de se nourir & de se re-
faire, estant humectée par le sang que contiennent les
petites venes ou arteres qui paroissent en sa superficie
exterieure; En sorte qu'aprés quelques temps, ses pores
estant deuenus plus estroits, les Esprits n'ont pas besoin
d'auoir tant de force qu'auparauant, pour la pouuoir
soutenir toute tenduë; Non plus que le vent n'a pas besoin d'estre si fort, pour enfler les voiles d'vn Nauire,
quand ils sont moüillez, que quand ils sont secs; Et cependant ces Esprits se trouuent estre plus forts, dautant
que le sang qui les produit, s'est purifié, en passant &
repassant plusieurs fois dans le cœur, ainsi qu'il a esté

CIII.
Comment
cette ma-
chine peut
s'éueiller
estant en-
dormie; &
au contra-
re,

cy-deſſus remarqué. D'où il ſuit que cette Machine ſe
doit naturellement réueiller de ſoy-meſme, aprés qu'elle a dormy aſſez long-temps ; Comme reciproquement
elle doit auſſi ſe rendormir, aprés auoir aſſez long-temps
veillé ; à cauſe que pendant la veille, la ſubſtance de ſon
cerueau eſt deſſechée, & ſes pores ſont élargis peu à peu,
par la continuelle action des Eſprits ; Et que cependant
venant à manger (ainſi qu'elle fait infailliblement de
temps en temps, ſi elle peut trouuer dequoy, pource
que la faim l'y excite) le ſuc des viandes qui ſe méle auec
ſon ſang le rend plus groſſier, & fait par conſequent
qu'il produit moins d'Eſprits.

CIV.
De ce qui
la peut ex-
citer à trop
dormir ou
à trop veil-
ler ; & des
ſuittes que
cela peut
auoir.

Ie ne m'arreſteray pas à vous dire, comment le bruit
& la douleur, & les autres actions qui meuuent auec
beaucoup de force les parties interieures de ſon cerueau,
par l'entremiſe des organes de ſes ſens ; Et comment la
joye & la colere, & les autres paſſions qui agitent beaucoup ſes Eſprits ; Et comment la ſechereſſe de l'air, qui
rend ſon ſang plus ſubtil, & choſes ſemblables, la peuuent empeſcher de dormir ; Ny comment au contraire,
le ſilence, la triſteſſe, l'humidité de l'air, & choſes ſemblables l'y inuitent ; Ny comment vne grande perte de
ſang, le trop ieuſner, le trop boire, & autres tels excés,
qui ont en ſoy quelque choſe qui augmente, & quelque
choſe qui diminüe la force de ſes Eſprits, peuuent ſelon
ſes diuers temperamens, la faire ou trop veiller, ou trop
dormir ; Ny comment par l'excés de la veille ſon cerueau ſe peut affoiblir, & par l'excés du ſommeil s'appe-
ſantir, & ainſi deuenir ſemblable à celuy d'vn homme
inſenſé, ou d'vn ſtupide ; ny vne infinité d'autres telles
choſes ;

ᴧchoſes ; dautant qu'elles me ſemblent pouuoir toutes
aſſez facilement eſtre deduites de celles que i'ay icy ex-
pliquées.

Or auant que ie paſſe à la deſcription de l'Ame Raiſonnable, ie deſire encore que vous faſſiez vn peu de reflexion, ſur tout ce que ie viens de dire de cette Machine ; Et que vous conſideriez premierement, que ie n'ay
ſuppoſé en elle aucuns organes, ny aucuns reſſorts, qui
ne ſoient tels, qu'on ſe peut tres aiſement perſuader,
qu'il y en a de tout ſemblables, tant en nous, que meſme auſſi en pluſieurs animaux ſans raiſon. Car pour ceux
qui peuuent eſtre clairement apperceus de la veüe, les
Anatomiſtes les y ont deſia tous remarquez ; Et quant à
ce que i'ay dit de la façon que les arteres apportent les
Eſprits au dedans de la teſte, & de la difference qui eſt
entre la ſuperficie interieure du cerueau & le milieu de
ſa ſubſtance, ils en pourront auſſi voir à l'œil aſſez d'in-
dices pour n'en pouuoir douter, s'ils y regardent vn peu
de prés. Ils ne pourront non plus douter de ces petites
portes, ou valvules, que i'ay miſes dans les nerfs aux entrées de chaque muſcle, s'ils prennent garde que la nature en a formé generalement en tous les endroits de
nos corps, par où il entre d'ordinaire quelque matiere
qui peut tendre à en reſſortir ; comme aux entrées du
cœur, du fiel, de la gorge, des plus larges boyaux, & aux
principales diuiſions de toutes les venes. Ils ne ſçau-
roient auſſi rien imaginer de plus vray-ſemblable, touchant le cerueau, que de dire qu'il eſt compoſé de pluſieurs petits filets diuerſement entrelacez, veu que toutes les peaux & toutes les chairs paroiſſent ainſi com-

C V.
Reflexion
ſur tout ce
qui a eſté
dit de cette
machine.

O

pofées de plufieurs fibres ou filets, & qu'on remarque le mefme en toutes les plantes; En forte que c'eft vne proprieté, qui femble commune à tous les Corps qui peuuent croiftre & fe nourrir par l'vnion & la ionction des petites parties des autres Corps. Enfin pour le refte des chofes que i'ay fuppofées , & qui ne peuuent eftre apperceües par aucun fens, elles font toutes fi fimples & fi communes, & mefme en fi petit nombre , que fi vous les comparez auec la diuerfe compofition , & le merueilleux artifice, qui paroift en la ftructure des or- ganes qui font vifibles, vous aurez bien plus de fuiet de penfer , que i'en ay obmis plufieurs qui font en nous, que non pas que i'en aye fuppofé aucune qui n'y foit point. Et fçachant que la Nature agit toufiours par les moyens qui font les plus faciles de tous , & les plus fim- ples , vous ne iugerez peut-eftre pas qu'il foit poffible d'en trouuer de plus femblables à ceux dont elle fe fert, que ceux qui font icy propofez.

CVI.
Que toutes les fonctiós qui luy ont efté attri- buées font des fuittes de la dif- pofition de fes orga- nes.

Ie defire que vous confideriez aprés cela, que toutes les fonctions que i'ay attribuées à cette Machine, com- me la digeftion des viandes, le battement du cœur & des arteres, la nourriture & la croiffance des membres, la refpiration , la veille & le fommeil; la reception de la lumiere, des fons, des odeurs, des goufts, de la chaleur, & de telles autres qualitez, dans les organes des fens ex- terieurs; l'impreffion de leurs idées dans l'organe du fens commun & de l'imagination; la retention ou l'em- prainte de ces idées dans la Memoire; les mouuemens interieurs des Appetits,& des Paffions;Et enfin les mou- uemens exterieurs de tous les Membres , qui fuiuent fi à

propos, tant des actions des objets qui ſe preſentent aux
ſens, que des paſſions, & des impreſſions qui ſe rencontrent dans la Memoire, qu'ils imitent le plus parfaitement qu'il eſt poſſible ceux d'vn vray homme; Ie deſire,
dis-ie, que vous conſideriez que ces fonctions ſuiuent
toutes naturellement en cette Machine, de la ſeule diſpoſition de ſes organes; ne plus ne moins que font les
mouuemens d'vne horloge, ou autre automate, de celle de ſes contrepoids & de ſes roües; En ſorte qu'il ne
faut point à leur occaſion conceuoir en elle aucune autre Ame vegetatiue, ny ſenſitiue, ny aucun autre prin-
cipe de mouuement & de vie, que ſon ſang & ſes Eſprits,
agitez par la chaleur du feu qui brûle continuellement
dans ſon cœur, & qui n'eſt point d'autre Nature que
tous les feux qui ſont dans les Corps Inanimez.

F I N.

INDEX

Absorption: from blood, *xxxvi*, 15; from intestine, 8; from stomach, 73
Accommodation of eye, 53, 56, 67n. *See also* Distance perception
Acids: in digestion, 6n; taste, 41
Acuity, visual, 55–58
Adam, Charles, and Paul Tannery, *xi*, *xvii*, *xliv*, 1n, 55n
Aging, 16
Air: as element or form of terrestrial matter, 2n, 3n; in hearing, 45n
Alchemy, 12n, 76
Alcmaeon, 18n
Alhazen (Hasan ibn Hasan), 51n, 68n
Anatomical Excerpts, *xviii*, *xlvi*, 16n, 19n, 71n
Ancient Medicine, 42n
Anger, 106
Angot, Charles, *xxiv*, *xliv*
Animism, *xxviii*. *See also* Soul
Anscombe, E., and P. T. Geach, *xvi*
Aorta, 13
Appetite, natural, 42n
Aqua fortis, 6, 68–69
Aqueous humor, 50, 51
Aquinas, St. Thomas, 104n
Aristotle, *xvii*; on automata, 4n, 5; on digestion, 6n; on the heart, 10n, 76n; on breathing, 32n; on pain and pleasure, 38n; on temperature sensations, 39n; on the common sense, 40n, 88n; on taste, 41n; on hearing, 46n; on dreams, 82n; on memory, 88n, 90n; on the soul, 114n
Arteries: pulmonary, 10, 11, 13; structure and function, 15–16; of brain, 17–20; of reproductive organs, 18
Aselli, Gaspare, 8n
Assimilation, *xxxvi*, 6n, 14, 15, 16

Association, *see* Memory
Atomism, 5n
Augustine, St., 2n
Auricles, 13n
Automata, 2n, 3n, 4, 22, 113
Automatic responses, *see* Reflexes

Bacon, Francis, *xxvii*
Baillet, Adrien, *xiii*, *xxiv*, *xlvi*
Bartholin, Caspar, *xvii*; on the lymphatics, 8n; on nerves, 24n, 34n; on the common sense, 40n; on smell, 43n; on memory, 88n
Bauhin, Caspar, *xviii*; on nerves, 24n, 30n; on the common sense, 40n, 88n; on taste, 42n; on hearing, 46n; on the pineal gland, 87n; on memory, 88n
Beeckman, Isaac, 6n
Berg, A., 8n
Beverwick, Johan van, 10n
Bichat, F. X., 77n
Bile, *see* Gall bladder
Binocular vision, 62–66
Blood: formation, *xxxvii*, 8n, 9; flow, 9–21; as distributor of heat, 12n; contents, 16–18; in emotions, 106. *See also* Circulation
Boas, Marie, 3n
Bonus, Petrus, 12n
Bordeu, Théophile, 77n
Boring, E. G., *xv*, 56n, 57n
Brain: flow of spirits through, *xxxvii*, 20–23, 27, 35–36, 37n, 71–72, 79, 80n, 81–111; blood supply, 19, 20; relation to organization, 22–23, 34, 36–37, 59–60, 64–65, 71–72, 77–112; membranes, 23n, 44; drainage, 44; in sleep, 82, 108; in dreaming, 109
Breathing, 12, 30–31

227

Cantelli, Gianfranco, *xiin*, *xvin*, *xlv*
Chanut, Pierre, 71n
Chapuis, A., 4n
Character traits, 72–73
Chemical change, 5n, 9n
Choanoid muscle, 53n
Choroid coat, of eye, 50
Choroid plexus, 19, 20, 50, 80, 86n
Church, Descartes and, 1n
Chyle and chyme, 8n, 9, 75n
Ciliary body, 53n, 55
Circulation, of blood: *xxxvi*, *xl*, 15, 16; pulmonary, 10–11, 13; fetal, 13
Clerselier, Claude, *xxiv*, *xxxiv*
Clock, body compared to, 22, 113
Cognition, *xxviii*, 12n, 60n, 85, 86, 88n. See also Imagination; Sensation; Volition
Cohen, I. Bernard, *xi–xvi*
Coimbran commentators, *xxxiii*, 72n
Colombo (Columbus), Realdo, *xviii*; on the lungs, 13n, 74n; on smell, 43n; on the eye, 52n, 53n; on animal spirits, 80n, 86n
Color theory, 53n, 59
Common sense, the, 40, 68n, 86, 87, 88n
Compendium of Music (*Compendium musicae*), 48n
Compounds, 3n
Conarium, *see* Pineal gland
Concoction, *see* Digestion
Condensation and rarefaction, 11, 13–15
Conduction, nervous, *xxxviii*, 21, 23n, 24, 33–34, 46, 83–84
Constitution, psychological, 72–76
Coordination, neuromuscular, 104
Copernicus, Nikolaus, *xiii*
Cornea, 51, 57n, 59n
Cornford, F. M., 47n
Corpuscles and corpuscularism, *see* Particles
Coughing, 33
Courage, 72, 106
Cranial cavity, 44n
Cribriform plate, 43n, 44, 45n
Crombie, A. C., *xlvi*, 5n, 47n, 52n, 55n, 63n
Crooke, Helkiah, *xviii*, 87n
Crystalline humor, *see* Lens

de Caus, Salomon, *frontispiece*, 68n
Defecation, 33
De Lacy, Phillip, 24n
Democritus, 3n

Descartes, René: biographical notes, *xiii*, *xxiii*; medical goals, *xiv*; philosophical ideas, *xv*, *xxx–xxxi*; biological works, *xviii–xix*; physiological theories, *xxvi–xxxiii*; historical position, *xxviii–xxx*; cosmological theories, *xxx*; sources, *xxxi–xxxiii*; physics of, *xxxv–xxxvi*
Description of the Body (*La description du corps humain*), *xviii*, *xxxix–xli*, 10n, 12n, 13n, 16n, 19n, 114–115
Development, *see* Embryogenesis; Growth
Diffusion, 8, 17
Digestion, *xxxvii*, 5–8, 17
Dijksterhuis, J. J., 3n, 114n
Dioptrics (*La dioptrique*), *xii*, *xlv*, 24n, 36n, 49n, 52, 56, 57n, 60n, 64n, 68n, 83n, 85n, 87n, 99
Discourse on Method (*Discours de la méthode*), *xii*, *xlv*, 1n, 10n, 12n, 13n
Discrimination, sensory, 33, 40n, 41, 60n, 85, 86, 102
Distance perception, 56, 57n, 59–68, 94
Dizziness, 80
Dreams, 82; symbolic dream of Descartes, *xl*
Dualism: of body and spirit, *xxvii*; of body and soul, *xxxi*, 1, 2n, 22, 36n, 113–114
Ductus arteriosus, 13
Dugas, René, *xin*
Dura mater, 23, 24n

Ear, 45–47
Earth, as element, 2, 3n
Eaton, R. M., *xvi*
Egestion, 33
Elements, 2n, 3n
Elizabeth (Princess of Palatine), *xxxi*, *xxxix*, 68n, 105n
Embryogenesis, *xl*, 7n, 12n, 19n
Emotion(s), 68n, 70, 105–106
Empedocles: on respiration, 32n; on psychological traits, 72n
Epicurus, Leucippus: atomic theory, 4n; on taste, 41n
Epigenesis, 19n
Epiglottis, 33
Epistemology of Descartes, *see* Knowledge, Cartesian theory of
Ethmoid bone, *see* Cribriform plate
Evaporation, 11n
Excretion, 33
Eye: *xxxix*, 49–68; muscles, 24–28, 53n, 62n; pigments, 51, 55. *See also* Aque-

ous humor; Vision; Vitreous humor; individual parts of eye
Eyelids, 29

Fabricius (Fabrizzi), Hieronymus, *xix*; on hearing, 46n; on vision, 54n
Faculties, natural, *xxviii*, 11n, 30n, 34n, 51n
Falloppio (Falloppius), Gabriello, *xix*, 30n
Fantasy, 88n, 96. *See also* Imagination
Fermentation, 12n; in digestion, 5–7; in blood formation, 9; in the heart, 11n, 14; in reproduction, 19n
Fernel, Jean, *xx*; on digestion, 6n, 7n; on fibers, 8n; on blood formation, 9n; on aging, 16n; on nerves, 30n; on breathing, 32n; on the common sense, 40n; on taste, 42n; on the natural appetite, 42n; on air reaching the brain, 44n; on memory, 88n
Fetus: development, *xl*; heart, 12–13; imprinting of, 87. *See also* Embryogenesis
Fiber(s) and fibrils, as constituents: of body solids, 7, 15, 16; of nerves, 24, 36; of the brain, 78–79
Filtration, 8, 17, 20
Fire: as element, 2n, 3n; nature of, 9n. *See also* Fermentation; Heart; Heat
Fixation, visual, 98–100
Fluid dynamics, 17, 18, 20
Foramen ovale, 13
Formation du foetus (alternate title for *Description du corps humain*), see *Description of the Body*
Fountains, *frontispiece*, 4, 21–22
Freedom, *see* Volition

Galen, *xxxii*; on natural faculties, *xxviii*; on digestion, 6n, 7n, 17n; on fibers, 7n; on absorption and blood, 8n, 9n, 13n, 20n; on the liver, 8n, 9n; on the heart, 11n, 13n, 29n; on the gall bladder and spleen, 17n, 75n; on respiration and breathing, 19n, 31n, 32n; on the pineal gland, 19n, 20n, 86n; on nerve structure and function, 23n, 24n; on muscle action, 25n, 26n; on the eye, 29n, 50n, 51n; on swallowing, 33n; on the senses, 34n, 39n; on pain, 38n; on temperature, 39n; on taste, 41n, 42n; on smell, 43n, 45n; on brain drainage, 44n; on hearing, 45n; on hunger, 69; on joy and

sadness, 70n, 71n; on psychology, 72n, 73n; on humors, 75n; on the heat of the heart, 76n; on spirits, 80n; on instincts, 104n
Galileo, D., *xii*
Gall bladder, 16–17, 30, 75
Gélis, E., 4n
Generation, *see* Reproduction
Generation (Primae cogitationes circa generationem animalium), *xviii*, *xlvi*, 18n
Georges-Berthier, Auguste, *xvn*, *xxxii*, *xlvii*, 8n, 10n, 11n
Gibieuf, Guillaume, *xxiii*
Gilson, E., *xx*, *xxxii*, *xxxiii*, *xlvii*, 11n, 19n, 72n, 89n, 108n, 111n
Girard, Theodore, *xxv*, *xliv*
God as Creator, 4
Greek influence on Descartes, *xxvi*
Growth, 15, 16

Haldane, E. S., *xiv*, *xlvii*
Hall, T. S., *xxviiin*, *xlvii*, 5n, 37n, 114n
Haller, Albrecht von, 57n, 77n
Harmony, *see* Music
Harvey, William, *xx*, *xxi*, *xxix*, *xl*; on absorption, 8n; on the heat of the heart, 10n, 74n; on rarefaction and condensation, 10n, 11n; on respiration, 74n; on spirits, 80n
Hasan ibn Hasan, *see* Alhazen
Hearing, 45–48
Heart: heat of, *xxviii*, *xxxvi*, 7n, 9–11, 12n, 76, 113; structure and function, *xxxvi*, 10–14, 70, 74–76
Heat: as movement of particles, *xxviii*; and life, *xxxvi*, 12n, 113; of the heart, *see* Heart; chemical source, 7, 9n; as fire without light, 9; as instrument of the soul, 12n; and cold, 39. *See also* Fermentation; Fire
Helmont, J. B. van: on digestion, 6n; on blood formation, 9n
Hippocrates, 42n, 70n, 72n
Homme, l' see *Traité de l'homme*
Humors, *xxvii*, 8n, 70n; as psychological traits, 107; crystalline, *see* Lens
Hunger, 68–69
Huyghens, Christian, 89n

Iatrochemistry and iatromechanics, 5n
Ideas, nature and basis, 41n, 80n, 84, 86–88, 92n, 94, 96n, 108
Image, visual: *xxxix*, 52; formation, 54, 57, 58n, 60–62, 84–85; illusion, 63–

68; projection onto brain, 84, 85n. *See also* Refraction
Imagination, 80, 86, 87
Imprinting, *see* Memory
Impulse, nervous, *see* Conduction
Inertia, *xxxvi*, 18
Infundibulum, 44n
Insomnia, 111–112
Instinct, 104n
Interpretation, Cartesian practice, *xxx*, *xl*, 4, 5, 80n, 112–113
Intestines, movement of food through, 7–8
Involuntary action, *see* Reflexes
Iris, 51, 55, 57

Joy and sadness, 70, 73, 105

Kargon, R. H., 3n
Keeling, S. V., *xiiin*
Kepler, Johann, *xxi*; on optics, *xxxix*, 52n, 54n, 56n; on conduction in nerves, 34n
Kidneys, 17
Kinesthesis, 63, 92–94
Knowledge, Cartesian theory of, *xxx*, *xl*, 1n, 5n
Koyré, Alexander, *xiin*
Kühn, K. G., *xxi*

Lacteals, 8n
La Forge, Louis de, *xli*, *xlvii*, 4n, 5n, 11n, 26n, 78n, 79n, 104n
Laughing, 106
Laurens, André du, *xxi*; on blood flow, 20; on nerves, 22n, 24n, 30n; on the common sense, 40n; on hearing, 46n; on the eye, 52n; on absorption, 74n; on animal spirits, 80n, 86n; on the pineal gland, 86n
Le Gras, Jacques and Nicolas, *xxiv–xxv*
Lens (crystalline humor), 50–57, 94n, 98
Lesky, Erna, 18n
Leucippus, 5n
Libavius, Andreas, 12n
Light: nature of, 49; intensity and perception, 60n. *See also* Color theory
Liver, *xxxvii*, 8, 9, 75n
Love, 72–73
Lucretius, *see* Epicurus
Lung, structure and function, 9–13, 74
Lymphatics, 8n

Massa, Niccolo, 43n
Mathematical reasoning, in Descartes, *xxix*, *xxx*, *xl*

Matter, theory of, 2n
May, Margaret T., *xxi*
Mechanism: in Descartes, *xv*, 1–4, 5n, 21–22, 80n, 104n, 112–113, 114n; in Harvey, *xxix*. *See also* Dualism; Soul
Meditations (Meditationes de prima philosophia), *xix*, 2n
Membranes, of brain, 23, 24n, 78n
Memory, 79n, 87–91, 96
Mersenne, Marin, *xiv*, *xxiii*, 6n, 8n, 12n, 13n, 17n, 38n, 40n, 47n, 49n, 87n, 89n
Mesaraic vein, 8n
Meteors (Les météores), *xii*
Microscopes, 5n
Mixed bodies, 3
Mnemonics, 90
More, Henry, 12n
Motion, nature and laws of, *xxxv–xxxvi*, 18
Motor nerve action, *xxxviii*, 22–23, 78. *See also* Muscle
Muscle(s): antagonistic action, *xxxviii*, 25–27, 29, 31–33; contraction, 21–25, 27–28; of eye and eyelid, 24–30, 53n; spirits in relation to, 25–28, 33, 36; tension in, 28–29; of breathing, 30–32; of swallowing, 32–34. *See also* Kinesthesis
Music, theory of, 47–48

Nasal secretions, 80–81
Nerve(s): structure and operation, *xxxvii*, 21, 23–25, 36, 78n; of the heart, 30, 68n, 70n, 75–76; olfactory, 43n; auditory, 45–46; optic, 53n; gastric plexes, 69n. *See also* Conduction
Newcastle, Marquis of, 18n, 83n
Newton, Isaac, *xii*
Nitric acid, 6n

Ogle, K. N., 63n
Olfaction, *see* Smell
Optic nerve, 53n. *See also* Retina
Optics, physical and physiological, 52–58
Organ, musical, brain compared to, 71
Orientation, muscular: of the eye, 28; of the limbs, 92–93

Pagel, Walter, 12n, 80n
Pain and pleasure, 37–38, 59, 107
Palate, drainage through, 44–45

Paracelsus: on spirits, *xxvii*; on digestion, 7n
Paré, Ambroise, *xxi*; on fibers, 8n; on aging, 16n; on emotions, 70n; on animal spirits, 80n, 86n; on the pineal gland, 86n; on memory and internal sensation, 88n; on sleep, 111n
Particles of matter, *xxvii, xxix, xxxvi*, 5, 17; elementary, 2n–3n; in digestion, 5–7; in chemical change, 7, 9n, 12n; in absorption, 8; and fire, 9n; in assimilation, 15–16; in blood, 16–20, 72–76; in the brain, 19–21, 60n, 80; and animal spirits, 19–21, 72–74, 80; in development, 19n; in muscles, 26; in touch, 33, 40; in taste, 41–43; in smell, 44. *See also* Spirits, animal
Partington, J. R., 3
Passions, *see* Emotions
Passions (Les passions de l'âme), *xix, xxxviii, xlv*, 2n, 10n, 12n
Peregrinus, Lelius, 105n
Pereira, Gomez, 2n
Perspective, theories of, 68
Perspiration, 17
Pharynx, 33n
Phlegm, 80–81
Pia mater, 23, 24n, 78n
Pica, 69
Piccolhomini, Arcangelo, *xxi*; on nerves, 24n, 30n; on the common sense, 40n, 88n; on smell, 43n; on spirits and the pineal gland, 86n; on memory, 88n; on sleep, 111n
Pineal gland, *xxxvii*, 19, 20, 37n, 40, 60n, 63n, 81, 85–88, 91–100, 109
Pirro, André, 47n
Plater, Felix, 52n
Plato: dualism of body and soul, 2n, 36n, 113n; on digestion, 6n; on reproduction, 18n; on respiration, 32n; on pain, 38n; on temperature, 39n; on taste, 41n; on hearing and sound, 47n; on memory, 88n, 90n
Pleasure, *see* Pain and pleasure
Plemp (Plempius), V.-F., 10n, 11n, 12n, 14n
Pneuma(ta) and pneumatology, *xxviii*, 80n. *See also* Spirits
Polyak, S. L., 52n
Pores (interfibrillar channels of the brain), *see* Brain
Porta, Giambattista de la, 52n
Position perception, *see* Distance perception
Price, D. J. de S., 4n

Principles (*Principia philosophiae*), *xii, xix, xlv*, 3n, 5n, 9n, 18n, 38n, 69n, 70n
Pulse, *xxxvi*, 14–15. *See also* Heart
Pupil, eye, 51, 54, 57

Rarefaction, *see* Condensation and rarefaction
Reflexes, *xxxviii*, 22, 33–35, 91–92, 101–102, 105, 108, 113
Refraction, *xxxix*, 50n, 51–58, 63–68. *See also* Image
Regius, Henricus, 2n, 9n, 11n, 114n
Regulae ad directionem ingenii, xix, 88n
Reproduction: *xl*, 18, 19n
Res cogitans and *res extensa, xxxi*, 113n
Respiration, 32n, 74n. *See also* Air; Breathing; Spirits
Rete mirabile, 80n
Retina, 50n, 51n, 54n, 55–59. *See also* Image; Refraction
Riolan, Jean, *xxii*; on fibers, 8n; on absorption, 74n
Rosenfield, L. C., *xlvii*
Rothschuh, K. E., *xvin, xxii, xlv*, 2n, 5n, 10n, 19n, 35n, 38n, 44n, 70n, 72n, 85n, 89n, 90n, 92n, 98n, 105n, 113n

Saliva, 17, 41–43
Scheiner, Christopher, *xxii*, 51n, 52n, 57n
Schneider, Conrad, 43n
Schuyl, Florentius, *xxiv, xli*, 2n, 3n, 77n
Schwann, Theodor, 8n
Sclera, 50
Scott, J. F., *xi, xlviii*
Sebba, Gregor, *xi, xlviii*
Secretion, 17
Semen, *xl*
Senility, *see* Aging
Sensation and sensory perception, *xxxviii*, 22, 37, 85n; discrimination, 33–34, 40, 41, 86, 102; internal, 68–73, 88n; physical basis, 84–88, 96, 100, 107. *See also* the individual senses; Common sense; Kinesthesis
Shape perception, 59–61, 85
Siegel, R. E., *xxii*, 8n, 22n, 34n, 74n, 86n
Sieve model of diffusion, 8, 17
Sight, *see* Vision
Sinus, sagittal and straight, 20
Size perception, 59, 62
Sleep, *xxxviii*, 81–82, 108–112
Smell, sense of, 43–45
Sneezing, 33, 79–80
Solmsen, F., 21n, 72n

Soul, *xxviii, xxxi, xxxviii, xli*, 1, 10n, 19n, 37–39, 41, 54, 60, 87n, 95–96, 113n–114n; interaction with body, 2n, 36n–37n, 88n; seat of, 36–37, 86, 95. *See also* Cognition; Emotion; Sensation; Volition
Sound, 46–48
Sphenoid bone, 44n
Spirits, animal, *xxviii, xxxviii*; in the blood, 17n; origin and nature, 19–21, 28, 80, 106–107; in the brain, 19n, 20–23, 27, 35–36, 37n, 71–72, 80n, 81–111; in nerves and muscles, 21, 25–28, 33, 36, 84–85; volatility, 28; Kepler on, 34n, 54n; in vision, 51n, 62n; varieties, 72, 73, 75, 80, 106–111; and psychological states, 72–73, 106
Spirits, natural and vital, *xxvii*, 32n, 80n
Spleen, 6n, 16–17, 30n, 70n
Stimulus and response, 34. *See also* Reflexes
Stockholm, *xxiv*
Stoic thinkers: on pneuma, *xxvii*; on vision, 62n
Swallowing, 32–34
Sweating, 17

Taste, sense of, 40–43
Temkin, Owsei, 80n
Temperature sensation, 39
Thirst, 69–70
Tingling, 38
Tissue, 77
Titillation, 42
Tongue, 41
Touch, sense of, 36–40
Tractatus De homine, xviii, xliii
Traité de la lumière, xii, xxiii-xxiv, 1n 18, 49
Traité de l'homme: planned and written, *xii, xxiii*; suppressed, *xiii, xxiv*; published and translated, *xxiv–xxv, xliii–xlv*; contents of original manuscript, *xxxiv–xlii*; illustrations, *xxxv, xli*

Traité du monde, xii, xv, xxiii
Traits, psychological, 72–73
Translations: of *l'Homme, xvi, xxiv, xliii–xlv*; of other works by Descartes, *xi–xii*
Transmission, *see* Conduction
Treatise of Light, see *Traité de la lumière*
Treatise of Man, see *Traité de l'homme*
Triangulation in vision, 62, 66
Tympanum, 46

Vacuum, 3n
Valves: of heart, 13–15; of veins, 15; of muscles, 26–28
Vaporization, *see* Condensation and rarefaction
Vartanian, Aram, *xlviii*
Veins, 15, 16; mesaraic, 8n; portal, 8–9, 74n; caval, 9, 13n; pulmonary, 10, 13; of brain, 20
Ventricles: of heart, 13; of brain, 19–20, 24. *See also* Brain
Vertigo, 80
Vesalius, Andreas, *xiv, xxii*; on fibers, 8n; on muscles, 24n, 30n; on nerves of the heart, 30n; on the common sense, 40n; on smell, 43n; on olfactory nerves, 44n; on the eye, 52n, 53n; on absorption, 73n; on the rete mirabile, 80n
Vision, *xxxviii, xxxix*, 49–68, 94, 98–100. *See also* Eye; Image
Vitreous humor, 50
Vives, Juan: on cognitive states, 88n, 105n; on sleep, 111n
Volition, 78n, 83, 108
Vorstius, Adolfus, 21n

Waite, A. E., 12n
Walking, 108
Walls, Gordon, 57n, 58n
Willis, Thomas, 44n
Witelo (Vitellius), 68n
World, The see *Traité du monde*